Quality in Higher Education
in the Caribbean

Quality in Higher Education in the Caribbean

EDITED BY

ANNA KASAFI PERKINS

 THE UNIVERSITY OF THE WEST INDIES PRESS
Jamaica • Barbados • Trinidad and Tobago

The University of the West Indies Press
7A Gibraltar Hall Road, Mona
Kingston 7, Jamaica
www.uwipress.com

© 2015 by Anna Kasafi Perkins

All rights reserved. Published 2015

A catalogue record of this book is available from the National Library of Jamaica.

ISBN: 978-976-640-512-0 (print)
      978-976-640-522-9 (Kindle)
      978-976-640-535-5 (ePub)

Cover design by Robert Harris
Typesetting by The Beget, India
Printed in the United States of America

In memory of Peter Whiteley

Contents

Acknowledgements	ix
Preface	xi
Introduction: Quality in Higher Education in the Caribbean *Anna Kasafi Perkins*	1

Part 1. Foundations

Chapter 1	Improving Learning through Effective Teaching *Anna-May Edwards-Henry*	21
Chapter 2	A Case for a Tertiary Education Council for the Caribbean Community *Halima-Saadia Kassim and E. Nigel Harris*	35
Chapter 3	The Quality/Financing Conundrum in Caribbean Higher Education *Kofi K. Nkrumah-Young*	49
Chapter 4	Science, Technology and Innovation: Entrepreneurial Universities for Caribbean Development *Patrick S. Dallas and G. Junior Virgo*	67

Part 2. External Explorations

Chapter 5	A Critical Look at the Legislative Framework for External Quality Assurance Agencies in CARICOM *Ruby S. Alleyne*	85
Chapter 6	Ethics and Quality Assurance: Purpose, Values and Principles *Anna Kasafi Perkins*	101

Chapter 7	Quality Assurance for Tertiary Level Technical and Vocational Education and Training for the Caribbean *Halden A. Morris*	119
Chapter 8	A Flexible Model for Quality Assurance in Technical and Vocational Education and Training in the Caribbean *Paulette J. Dunn-Pierre*	133

Part 3. Internal Issues

Chapter 9	A Summative Evaluation Model for Strategic Planning in Tertiary Education *John Gedeon*	149
Chapter 10	Internationalization, Cross-Border Tertiary Education and the Challenge for Quality: A Jamaican Perspective *Dameon A. Black*	171
Chapter 11	A Quality Scorecard Approach to Analysing Quality in Distance Online Education *Pamela Dottin*	181
Chapter 12	Massive(ly) Open Online Courses: Opportunity or Threat to Caribbean Higher Education? *Patrick Anglin*	193
Chapter 13	Building a Quality Institution of Higher Learning in a Small State: Issues, Considerations and Challenges *S. Joel Warrican*	209
Chapter 14	The University of the West Indies: Moving Quality to the Next Level *Sandra Gift*	225
Chapter 15	Details, Details, Details: Administrative Personnel and Quality Assurance *June Wheatley*	237
Chapter 16	Transforming Higher Education in the Caribbean: The Total Quality Management/Service Quality Model *Eduardo R. Ali*	255
Chapter 17	Future Directions for Quality Assurance in Higher Education *Anna Kasafi Perkins*	269

Contributors   275

Acknowledgements

Working on an edited volume epitomizes the collegial, team-based, values-laden approach towards higher education. The melding of numerous energies, efforts, shared talents and skills came together in a marvellous way to make this possible. There are many to thank, the contributors for sharing their intellectual and practical experiences as well as the commentators, who walked alongside individual contributors from conceptualizing to completion. I am grateful to Rohan Lewis, Elsa Leo-Rhynie, Alan Cobley, Alvin Wint, Carolyn Hayle, Glen Boyne, Donna Lyn Fatt, Zara Orane, Vivienne Roberts, Gloria Barrett-Sobers, Marcia Stewart, Dawn Barrett-Adams, Joseph Pereira, Densil Williams, Canute Thompson, Valda Alleyne, Robert Carroll, Charmaine McKenzie, Mervin Chisholm, Anthony Perry and Disraeli Hutton. Thanks too to those who began the journey but were unable to finish it for various reasons.

Thanks also to Linda Speth and her team at the University of the West Indies Press and the anonymous reviewers who believed in this work.

Preface

It is an honour to have been asked to provide a preface to this pioneering work on quality assurance in higher education in the Commonwealth Caribbean. The range and scope of the contributions brought together here so ably by Anna Kasafi Perkins are eloquent testimony to how far we have come in the development of a robust and vibrant indigenous quality assurance culture in higher education in our region over the past three decades. Several chapters detail efforts to embrace quality assurance as a core internal academic and administrative activity by institutions from community colleges to universities across the region, while others refer to the emergence of national accrediting bodies for higher education in several countries. The spread of the quality assurance culture to new areas, such as technical and vocational education and online education, is also documented. Hopes for the future – including ideas for new, more holistic models for quality assurance in higher education institutions, and the provision of a regional accreditation system supported by CARICOM – are also discussed. In the process, quality assurance is defined to mean, variously, "relevance", "value for money", "fitness for purpose", "accountability" and the use of "external benchmarking". As this collection shows, quality assurance in higher education can mean all of these and more.

If the formation of the University Council of Jamaica in 1987 could be said to have marked the beginning of a journey into a new era of quality in tertiary education for the region – the first of a number of milestones identified by Perkins in her introduction to this volume – it would be a mistake to suppose that we are nearing the final destination in that journey. Rather, as the various contributions to this collection suggest, we have arrived at a crossroads in the development of the quality assurance culture in the Commonwealth Caribbean.

One reason for this is that the providers in the tertiary education sector in our region continue to grow and multiply, driven by the insatiable aspiration of our people for educational advancement, on the one hand, and the new economic realities of the twenty-first century, on the other. To cut a long story short, tertiary education is more vital to the development of the people of the region than ever, while, simultaneously, the commodification of knowledge is turning the provision of further and higher education into big business, forcing public and private providers in the sector into competition with each other for the student's dollar as never before. To these new realities, we must add a third: in recent years the onset of severe financial crisis has cut deeply into government revenues and the disposable income of private citizens across the Commonwealth Caribbean, so that the means of funding higher education has become a pressing question for policymakers across the region. What role can quality assurance principles and mechanisms play in the region in this dynamic and potentially dangerous environment?

Another reason for suggesting that we stand at a crossroads in the development of a quality assurance culture in the Commonwealth Caribbean is to do with the ubiquitous influence of new technologies on all our lives. The inexorable rise of new information and communication technologies means that competition among institutions in higher education is now global, and many institutions are turning to online and blended delivery modes to contain unit costs and expand their student catchment into new geographical areas. Traditional institutions in our region that still rely primarily on face-to-face delivery of their courses and programmes are being challenged to demonstrate their continuing relevance in such circumstances. The long-term survival of at least some of these institutions will depend on their capacity to adapt to, and engage with, these new modalities in education. Although it is recognized that these new modes of delivery bring their own unique challenges from a quality assurance perspective, efforts in the region to develop appropriate mechanisms to measure the quality of these programmes or to establish regulatory bodies to accredit and set standards for them are still in their infancy. This is another area in which urgent action is required from quality assurance professionals and policymakers in the region.

The opportunities, as well as the dangers, arising from the new technologies are amply illustrated in the case of massive open online courses (MOOCs). MOOCs are providing previously undreamed of opportunities for access to higher education, and have been seized upon as an appropriate vehicle by some of the leading universities in the world. The direction this trend in higher education will take is still highly contested and remains unclear. However, it is not hard to imagine what could happen without the participation of indigenous institutions in the design and delivery of courses for the region, and without the necessary checks and balances that appropriate quality assurance systems would provide. Ultimately, this global process of "democratization" of access to higher education may tend to consolidate the power of a relatively small group of elite universities from the developed world, while marginalizing and undercutting less technologically resourced universities across the developing world. It would be sad indeed if the hard fought decolonization of higher education in the global South over the second half of the twentieth century were to be superseded by a technologically driven recolonization in the first half of the twenty-first century.

Over a decade ago, one of my predecessors in the Office of the Board for Undergraduate Studies (OBUS) at the University of the West Indies, Hilary Beckles, co-authored a book (with Anthony Perry and the late Peter Whiteley, also of OBUS) entitled *Brain Train: Quality Education and Caribbean Development* (2002). They styled their book "a twenty-first century manifesto". Surveying the landscape of Caribbean development as it appeared to them then, they declared: "Nothing short of an education revolution, located primarily in the higher level of the sector, will be adequate. Mass access to quality higher education – the 'Brain Train' – is the Caribbean's last chance to secure sustainable development" (p. vii).

If mass access to quality higher education seemed vital for Caribbean development *then*, it certainly is no less vital now, as we sit in the eye of a global economic, social and cultural storm. The key word there is "quality". In this collection, Perkins and her contributors challenge us to imbue the word "quality" in higher education with real meaning, and to act decisively to use quality assurance as a critical tool in the next phase of Caribbean development.

One final thought. I believe that one of the important messages conveyed by this book is that we need to begin thinking about, and to start planning for, quality assurance in higher education on a regional level in the Commonwealth Caribbean. In their contribution to this volume, Harris and Kassim offer one suggestion for the way forward – a tertiary education council for CARICOM to provide "improved coordination for long-term planning and overall development of the tertiary education system". Whether this particular suggestion gains traction remains to be seen, but the case for a body with formal oversight of the higher education system in the region at this stage in our development seems to me to be overwhelming.

It is my hope that the many ideas and perspectives presented in this book will inform a productive debate on the future direction of quality assurance in higher education in the Commonwealth Caribbean. Such a debate is a prerequisite for decisive action to build the capacity of tertiary institutions across the region if they are to serve the development of Caribbean people better in this era of rapid change.

Alan Cobley
The University of the West Indies
Cave Hill, Barbados

Introduction
Quality in Higher Education in the Caribbean

ANNA KASAFI PERKINS

The European University Association stated that "Quality assurance refers to a set of procedures adopted by higher education institutions, national education systems and international agencies through which quality is maintained and enhanced. Quality assurance is effective when it refers to the very core of the higher education activity and when its results are made public" (European University Association 2001). Quality assurance is different from accreditation, which the European University Association considers "as one possible outcome of quality assurance and defines it as a formal recognition of the fulfilment of minimum, publicly stated standards referring to the quality of a programme or an institution" (2001). Accreditation is an adequate mechanism for assuring minimum standards of education and, in some cases, can be seen as the first step towards assuring quality in higher education. Quality assurance in higher education has come of age in the Commonwealth Caribbean region.

How far quality assurance in higher education in the Caribbean has come is indicated by some important milestones: the twenty-fifth anniversary of the region's first national accrediting body, the University Council of Jamaica (founded 1987); the ninth anniversary of the Caribbean Area Network for Quality Assurance in Tertiary Education (CANQATE; founded 2004); the tenth anniversary of the Caribbean Accreditation Authority for Education in Medicine and other Health Professions (CAAM-HP; founded 2003); and the first *institutional* accreditation of the two universities: the University of Trinidad and Tobago (UTT) and the College of Science, Technology and the Applied Arts of Trinidad and Tobago in 2010. These were the first institutions accredited by the Accreditation Council of Trinidad and Tobago (ACTT). At the same time, quality assurance has become increasingly recognized as a professional practice undertaken by qualified Caribbean professionals requiring opportunities for training, reflective practice, research and international collaboration. All of these developments have kept pace with global trends and best practices in quality assurance in higher education (El-Khawas, DePietro-Jurand and Holm-Nielsen 1998).

Nonetheless, questions of relevance, accountability, access and transparency in tertiary education continue to arise (Beckles, Perry and Whiteley 2002; UWI 2006). Within the last three decades, these concerns have been addressed through the development of robust internal quality assurance processes within institutions of higher learning supported by national external quality assurance agencies (EQAAs), like the University Council of Jamaica and the ACTT (established 2005). Even so, several of the accreditation bodies in the region are still fairly embryonic; neither Belize nor St Lucia, for example, has functioning accreditation boards, in spite of relevant legislation being in place. Institutions with internal quality assurance systems continue to strive for enhancement, including the creation of dedicated quality assurance offices, policies and strategies, and encourage a quality culture in which a larger quality management framework is embedded.

The University of the West Indies

The regional university, the University of the West Indies (UWI), which was established under Royal Charter in 1948, has perhaps the most mature internal academic quality assurance system in the region. "[Indeed], the organisational structure of the University is designed with specific quality assurance objectives in mind" (UWI 2000/2001, 4). Nonetheless, UWI established a quality assurance unit (QAU) in 2001 under the aegis of the Board for Undergraduate Studies to keep abreast with wider developments. The Board for Undergraduate Studies, recognizing the centrality of the student to the mission of the UWI, called for: "[A] robust Quality Assurance and Quality Audit [now known as Evaluation] system at all levels of the operation of UWI. . . . [to] allow students (and, indeed, other stakeholders including parents, employers, governments) to be confident that qualifications from UWI continue to represent the product of a high-quality, student-focussed university" (UWI 2000, 3).

At the same time, other stakeholders are also taken account of in the process. "The intention of [the Office of the Board for Undergraduate Studies] OBUS [, the operational arm of BUS and the office within which the QAU was seated,] was to build on what already existed, improve the level of functioning where necessary and extend the systems" (Whiteley 1998, 8).

The QAU has responsibility for academic quality as part of an unfolding quality management system (QMS) for the University. The QAU contributes to policy and undertakes research, advocacy, education and scholarly publications in the area of assurance and enhancement. Officers also have an outreach presence and have provided consultancies with the Ministry of Education in Suriname, the University of Guyana (UG) and Hugh Wooding Law School, among others. In 2011, an external team reviewing the QAU commented that "the Quality Assurance Unit (QAU) has had a positive impact upon the operation of the UWI with respect to its management of quality and the enhancement of the student learning experience" (Review Team Report 2011, 2).

Accreditation of Professional Programmes

At the same time, UWI has had a long history of seeking international specialized accreditation for disciplines like engineering and medicine. This practice predates the formation of the QAU. Such professional disciplines need to retain the "accreditation of regulatory professional bodies, which conduct periodic intensive reaccreditation exercises to ensure that the programmes being offered meet international standards" (Leo-Rhynie and Hamilton 2008, 312). The UWI later spearheaded the formation of the regional accrediting body for medicine and other health professions (CAAM-HP), when the General Medical Council of the United Kingdom gave notice that after 2003 it would no longer be accrediting medical schools in the Commonwealth.

UWI's engineering degrees are accredited by agencies of the Engineering Council of the United Kingdom, which is a member of the Washington Accord, an international mutual recognition agreement signed in 1989 among the major English-speaking countries. Accreditation by a member of the Washington Accord is an indication of the high quality of UWI's degrees; however, it does not guarantee automatic recognition by other members of the Accord. Such automatic recognition is accorded to accreditation by a member from within its own national borders. At the time of the inauguration of the Washington Accord, the Commonwealth Caribbean was the only major English-speaking region that was not (and is still not) a member since it has no indigenous accreditation system for engineering and technology. The Accord decreed that "countries and regions should establish their own accreditation agencies of international standards and seek admission to the Washington Accord, rather than depend indefinitely on external accreditation agencies" (CACET n.d.). The regional Caribbean Accreditation Council for Engineering and Technology was officially established on November 26, 2009, in San Juan, Puerto Rico, at a meeting of members of the Caribbean engineering fraternity, a meeting which included UWI and the University of Technology, Jamaica (UTech), the national accreditation agencies in the region, academics from regional universities and representatives of the Caribbean Community (CARICOM). Since its establishment and formal recognition by CARICOM, the Caribbean Accreditation Council for Engineering and Technology has accredited thirteen engineering programmes.

Institutional Accreditation

Beginning with the St Augustine campus in 2011, all the campuses of the UWI have now been accredited by their national bodies. Mona was accredited by the University Council of Jamaica in 2012, Cave Hill and the Open Campus by Barbados Accreditation Council in 2013. (The Open Campus was also accredited by ACTT in July 2014 as part of the process of mutual recognition.) The long-discussed CARICOM regional accrediting body has yet to come into being, and so national accrediting agencies, through the efforts of CANQATE, are involved in discussions on mutual recognition, a move that is in keeping with global best practices, as is detailed in the International Network of Quality Assurance Agencies in Higher Education (INQAAHE) *Guidelines for Good Practice in External Quality Assurance Agencies* and the European Association for Quality Assurance in Higher Education 2002 Occasional Paper 4: *A Method for Mutual Recognition Experiences with a Method for Mutual Recognition of Quality Assurance Agencies*.

University of Trinidad and Tobago

In 2009, the UTT created an Office of Quality Assurance and Institutional Effectiveness with "primary responsibility for the design, development and implementation of a university-wide system for quality assurance and accreditation and an integrated approach to institutional research that is linked to strategic planning and evidence-based decision-making" (https://u.tt/); UTT, a publicly funded institution that was founded in 2004, was awarded institutional accreditation for seven years in 2010 by the ACTT. The ACTT accreditation report to UTT states the following:

> The team commends UTT for its achievements in creating the university over the last six years, for developing a clear and well supported mission, for integrating predecessor organizations into the University, and for the development of a strong academic and quality infrastructure. The successful expansion of the University has not only increased the numbers of people in higher education, it has widened access. A significant number of students told us that they had entered higher education only because of the relevance of the programmes offered by UTT; such people would otherwise have been lost to higher education, and their potential left unfulfilled. (https://u.tt/)

The UTT also has several specialized programmes that have been internationally accredited. These include: the Master of Science in Operational Maritime Management (2011–16), Bachelor of Science in Nautical Science (2011–16), Diploma in Maritime Operations – Engineering Option (2011–16), and Diploma in Maritime Operations – Navigation Option (2011–16). All these are accredited by the Institute of Marine Engineering, Science and Technology in the United Kingdom. Other degrees are accredited by the Energy Institute, United Kingdom.

The University of Guyana

The UG demonstrates the challenges inherent in developing a formalized quality assurance system in a Caribbean higher education institution. In 2010, under a previous vice chancellor, UG collaborated with the UWI in its quest to develop an internal quality assurance mechanism (*Stabroek News*, July 20, 2009). A series of workshops were conducted with select staff and administrators of UG by officers of the UWI QAU and one employee had a brief work shadowing visit with the QAU. The set-up of the UG QAU is estimated to cost G$8.8M, as seen in the 2011 budget. However, these plans are yet to be implemented. The quality assurance system as articulated by the current vice chancellor will include utilizing second examiners and external examiners for all programmes (*Kaiteur News*, September 24, 2013). Among the new proposals by the incumbent vice chancellor for restructuring the University are increasing lecturer-teaching hours to at least eighteen per week, in a bid to obtain "value for money" (ibid.). At the same time, the vice chancellor was clear that efforts had to be made to address low staff salary and compensation. He noted that "We cannot run a quality University using salaries that are not conducive to decent living" (ibid.). He also mentioned efforts to facilitate the upgrade of the university's infrastructure.

The impact of these proposed changes on the quality of UG programmes was recently called into question by the president of the University of Guyana Senior Staff Association. According to the president of the University of Guyana Senior Staff Association, the proposal to increase teaching hours is tantamount to downgrading UG to a "high school of arts and science" as it is only in high schools in Guyana that teachers carry such hours (*Kaiteur News*, September 24, 2013). Indeed, the current hours of faculty teaching are predicated on involvement in scholarship and outreach, the president detailed, while arguing that of fifty-six universities surveyed (including UWI), no lecturer taught twelve hours as is the current case with UG lecturers. One respondent to the online article in the *Kaiteur News* detailing the concerns of the president of the staff association stridently rejected the notion of external examiners as a quality assurance method. Indeed, the writer, who calls him- or herself, "Paid Piece Piece" declared, "And, to talk about outside examiners, is the biggest insult to the lecturers!! That used to be a colonial strategy done in the days of old!! Are we moving backwards or moving into the 21st century!!! Lecturers should strike if it will take outside examiners to decide whether a student passed a course or not!!" (ibid.). Clearly, the debate on the meaning of quality in higher education in the Caribbean is alive and acerbic. The history of colonialism colours even ideas about the nature and purpose of quality assurance and this has to be taken into account in developing a quality assurance culture in Caribbean higher education institutions.

Anton de Kom University of Suriname

The Anton de Kom University of Suriname is the only university in Suriname. Since its founding in 1966, the university has gone through several reorganizations, including the introduction of a four year bachelor's (1983), of which the last six months was intended for internships, and a master's (2008–9). In 2008, the National Agency for Accreditation was established and all higher education institutions were expected to have applied for accreditation no later than 1 October 2010. The National Agency for Accreditation offers programme accreditation; institutional accreditation is not a possibility under the law. In response, the Anton de Kom University of Suriname dubbed 2010 "the Year of Accreditation". Anton de Kom prepared for accreditation by establishing an Accreditation Steering Committee and a dedicated web page (http://www.uvs.edu/). In April 2013, the accreditation criteria were approved by the minister of education and the first programme (the Anton de Kom University of Suriname master's in petroleum geology) was accredited for six years. The Anton de Kom University of Suriname intends to apply for accreditation of remaining programmes within three years.

The university has recently reorganized to provide support for quality assurance, education and research development in a new office with a new director. The former Institute of Quality Assurance is integrated in the new office. They are developing an internal quality assurance process based on the European Foundation for Quality Management and the external quality assurance (EQA) process from the National Agency for Accreditation (Pieters in email to author, 2014).

Professionalizing Quality Assurance

Quality assurance has become an increasingly professional undertaking, with a singular identity, body of knowledge and trained practitioners. Professionals involved in designing and implementing quality assurance systems in the region can claim membership in both local and international professional organizations boasting codes of good practice, professional development activities, research and publications.

INQAAHE

The CANQATE is a subnetwork of the INQAAHE. Both INQAAHE and CANQATE hold regular international conferences and provide specific training for quality assurance practitioners. INQAAHE developed its online graduate programme in quality assurance in response to: "The massive increase in external and internal quality assurance (QA) activity over recent decades, together with the associated thinking about it, [which] have created a new profession that requires a structured academic discipline and programs to educate quality assurance professionals, stimulate research, and produce new initiatives" (http://www.inqaahe.org/).

The LH Martin Institute for Higher Education Leadership and Management at the University of Melbourne, Australia, delivers a one-year, part-time graduate certificate in quality assurance for quality assurance practitioners in tertiary education. It was developed in close collaboration with INQAAHE and is awarded by the University of Melbourne through its Melbourne Graduate School of Education. Quality assurance professionals from the region have been certified through this programme.

CANQATE

Closer to home, CANQATE, with the support of UNESCO and other partners, has provided online and face-to-face training opportunities for Caribbean professionals. In 2012, the CANQATE capacity building programme on EQA, which was adapted for the region from UNESCO's International Institute for Educational Planning EQA training programme, turned out twenty graduates hailing from across the region (Coordinator's Report 2012). CANQATE also provides scholarships to assist members to attend the INQAAHE conferences held every two years.

ACTI and CARICOM

Other regional associations, like the Association of Caribbean Tertiary Institutions (ACTI), contribute towards "creat[ing] within the Caribbean region of a learning Society which offers opportunity, and strives for quality and harmony in the diverse tertiary education environment" (www.acticarib.org). ACTI regularly undertakes projects, such as the 2012 Harmonization and Articulation of Associate Degree Programmes to Be Offered by Regional National and Community Colleges, which strengthen the CARICOM goal of harmonization of certification and accreditation processes to facilitate the mobility of the workforce. ACTI, ACTT and the Association of Caribbean Higher Education Administrators sponsor regular international conferences for the higher education community that focus on areas such as quality assurance and enhancement while taking account of global trends.

The CARICOM Secretariat has also undertaken technical capacity building programmes aimed at national accreditation bodies. In 2011, for example, the Technical Action Services Unit of the CARICOM Secretariat, with the support of the European Union, coordinated a capacity building project targeting the accrediting agencies of Antigua and Barbuda, Barbados, Belize, Dominica, Grenada, Guyana, St Lucia, St Kitts and Nevis, St Vincent and the Grenadines, and Suriname. The key activities of the capacity building programme included training of assessors in the techniques and practices related to quality evaluation; increasing awareness among key stakeholders on core issues related to quality assurance; and training for peer and external evaluators (CARICOM 2011).

Quality in Higher Education in the Caribbean

This volume, *Quality in Higher Education in the Caribbean*, is an attempt to recognize and to critically assess these seminal developments in the quality assurance arena in the Caribbean. In so doing, *Quality in Higher Education in the Caribbean* brings together the voices of quality assurance and other higher education practitioners from across the higher education sector in the English-speaking Caribbean. The contributors use their experiences and research to reflect on some of the key issues which have implications for quality, for example, the financing of tertiary education, assessing student learning, the impact of the internationalization of higher education on the indigenous systems of tertiary education, accreditation and strategic planning, among others.

The collection is divided into three related parts. Part 1, "Foundations", deals with select issues in higher education that impact quality assurance, beginning with teaching and learning. The main activity at which quality assurance activities are aimed is learning or student achievement and how effectively this is transferred to life and work situations (Knapper 2006). Indeed, studies have shown that the implementation of quality assurance systems "have caused academic institutions to give greater attention to issues of effective teaching and learning" (El-Khawas, DePietro-Jurand and Holm-Nielsen 1998, 7). Teaching in all its forms is the main mechanism through which learning is facilitated. Former president of the UTech, Alfred Sangster, in a discussion of education and training in the development process, expressed the need for "a critical review of the methods of learning with an increased use of educational technology" (1994, 207). Anna-May Edwards-Henry of the Centre for Excellence in Teaching and Learning, UWI St Augustine, sets the stage for the discussion on quality by establishing the centrality of learning in the teaching enterprise in chapter 1. She critiques teaching as a teacher-centred activity for which lecturing is the most popular mode. (Edwards-Henry deliberately uses the term "teacher" as opposed to lecturer or instructor to emphasize the role of teaching.) In calling for a rethinking of teaching, Edwards-Henry explores the six determinants of learning with a view to contributing to more effective teaching at the tertiary level in the Caribbean. Equipping teachers to meet the dynamism of the learning enterprise is necessary to "attain high-quality teaching and sustained learning improvement".

Chapter 2 sees the vice chancellor of the UWI E. Nigel Harris and senior planning officer of the St Augustine campus, Halima-Sa'adia Kassim, making the case for a tertiary education council for the CARICOM region. Kassim and Harris, in describing the fragmented, diverse, under-resourced nature of higher education in the CARICOM, approach the same issues highlighted by Leo-Rhynie and Hamilton (2008, 311), who call for "a quality driven tertiary education system [that] . . . demonstrate(s) flexibility and responsiveness to the requirements of local communities and national as well as regional needs". The system as it currently exists suffers from a set of challenges that makes it imperative to address issues of duplication and redundancy, quality assurance, linkages between education programmes and societal needs, fragmentation, insufficient resources and inadequate cooperation that will enhance coordination, coherence and sustainability. The proposed CARICOM Tertiary Education Council is seen as the mechanism to achieve economies of scale and scope in the sector, reduce disparities in access and equity, improve data sharing and greater transfer of knowledge, skills and technologies, and foster more relevant and beneficial research in all areas of Caribbean society and economy. The authors explore the scale and scope of the tertiary education sector, the relevance and value of a tertiary education council and examine the governance and financing arrangements associated with the development of that council.

As Harris and Kassim highlight in chapter 2, financing has stymied the development of the tertiary education council. At the same time, CARICOM, in deliberating about a tertiary education system, wrestled with linking performance to funding. Similarly, the question of linking performance to state funding of higher education is alive and well, as Kofi Nkrumah-Young, the vice president of Planning and Operations at the UTech, demonstrates in an analysis of the current nexus between the national quality assurance mechanisms and the funding model utilized in Jamaica, Barbados, and Trinidad and Tobago. His key question is, "How is quality linked to the funding model?" Are the funding models employed by the select Caribbean governments impacting on the quality of the higher education provided? Using Orr's trajectory to analyse the relationship between funding of teaching and quality assurance, Nkrumah-Young reveals that for Trinidad and Tobago, the quality assurance mechanism (registration), being compulsory, correlates to the resource allocation mechanism of discretionary budgeting. In the Jamaican and Barbadian cases, however, state funding is not linked to the quality assurance mechanism (registration or accreditation are not compulsory) thus those governments do not have systems in place to communicate the quality effect of their funding. Nkrumah-Young concludes that the quest for efficient use of resources, however, suggests that there ought to be a more direct link between the method of allocating state funds to higher and tertiary institutions and the method of providing information on the impact of those funds on the quality of teaching as well as research output.

Nkrumah-Young's analysis must be viewed against unfolding events in the region. In August 2013, the government of Barbados announced in the budget debate that, beginning from academic year 2014, all students will pay their tuition at university. Education in Barbados has been essentially free up to university level since its independence in 1966. This has been crucial to the island's development; literacy rates of approximately 98 per cent have made many Barbadians proud. The government's recent decision has spawned much debate, encompassing issues of education policy and debt management. The Barbadian government plans to look to means-testing and a revamped students' revolving loan programme to ensure that deserving students can still have access to a university education.

Since the 1980s, Jamaica instituted a form of cost sharing in higher education – the government pays 80 per cent of the tuition costs and students pay 20 per cent. The matter concerning how much should be paid by students in Jamaica is by no means settled. Many students turn to student loans to finance their education. The student loan system has been rife with challenges, not the least of which is insufficient funds to cover demand and high delinquency rates (Johnson 2013). As recent as September 2013, Jamaica's House of Representatives solicited written opinions on how the country should pay for tertiary education. Members of the public were invited to give their views on a motion relating to the funding of tertiary education in Jamaica.

In chapter 4, industry leader and policymaker Patrick Dallas and UTech engineering lecturer G. Junior Virgo venture into the key area of scientific and technological advances and the impact of the higher education sector in that area. They maintain that, given the science and technology revolution that continues to transform our world, the Caribbean should also expect its universities, as leaders in science and technology education, research, and innovation, to play an important role in economic development by transforming into entrepreneurial universities in partnership with government and industry. Jamaica's former ambassador to Washington, Audrey Marks, in accepting the honorary degree at Northern Caribbean University, summed up the importance of the role of the university in science, technology and innovation: "To a large extent, Jamaica's future will be decided by the extent to which institutions like Northern Caribbean University establishes a tradition of innovation and scientific enquiry and encourages Jamaicans from all walks of life to correct the weak research and innovation culture which presently prevails" (*Jamaica Observer*, August 27, 2012). Using examples mainly from Jamaica, Dallas and Virgo demonstrate the ways in which this weak research and innovation culture is being strengthened and the impact on development. They explore the nature of the relationship between the three central players in the science, technology and innovation nexus – the university, the state and industry. The entrepreneurial university that engages state and industry partners effectively is uniquely poised to both set the agenda for science, technology and innovation, and develop innovations that contribute to the national development agenda.

Clearly, quality assurance in science education and research can support the transformation of our universities into entrepreneurial institutions through such initiatives as setting standards for relevant curricula, qualified and motivated staff, appropriate partnerships with both government and industry (as Dallas and Virgo argue), and processes to support, encourage and utilize innovation by students and faculty. Constant investment in facilities such as labs and state-of-the-art equipment will provide the tools for science and technology, as well as create a larger entrepreneurial culture.

Part 2, "External Explorations", looks at quality-related matters within the EQA context. Ruby Alleyne, vice-president, Quality Assurance and Institutional Effectiveness at the UTT, outlines the process that led to the development of national systems for quality assurance and accreditation in CARICOM. In chapter 5, she brings together in one place the various legislations that undergird the work of such agencies. She provides a thorough analysis that identifies points of convergence and divergence among the legislation. Such an analysis is increasingly important as these Caribbean EQAAs move towards mutual recognition of national accreditation decisions with the absence of a regional authority. Alleyne is clear that issues of standards and quality are important and should be addressed in an effort to ensure comparability and coherence. Only then can the vaunted mobility of workers across the region, a key aim of the Caribbean Common Market, be realized.

One of the less explored areas in the quality assurance movement is the ethics of quality assurance. Higher education is significantly impacted by corruption and other unethical practices. The extent of this problem is revealed in 2013 Transparency International Survey, which focused on education. This is the first such report by the organization focusing on the problem in education. (Elizabeth Buckner [2013] has critiqued the report for not focusing on Arab universities. Similarly, there was no attention to the Commonwealth Caribbean in the report.) Clearly, corruption and misconduct in education have significant costs; they undermine the teaching and learning mission of tertiary education. Corruption costs real money and undermines the legitimacy and meritocracy that universities claim to provide. In chapter 6, Anna Kasafi Perkins, senior programme officer QAU, UWI, serving the Mona campus, introduces the role and purpose of quality assurance in ensuring ethics in higher education. Perkins explores the commitment related to ethics espoused by selected Caribbean accreditation agencies – the University Council of Jamaica, CAAM-HP, Barbados Accreditation Council and ACTT – in their core values and mission. In so doing, she clarifies the notion that quality assurance itself is expected to be an ethical practice and is undergirded by the same values it ensures. Perkins's chapter further discusses the kinds of ethical lapses that have taken place in the process of accreditation. It takes account of examples of alleged or potential ethical issues related to three EQAAs, two of which are drawn from the Caribbean region. It closes by outlining briefly an ethics framework for EQAAs that starts with values identification and ends with a process of self-evaluation.

The specialized quality assurance in technical and vocational education and training (TVET) is explored in chapters 7 and 8. These two chapters trace the terrain for TVET in a fashion that opens up the important linkages between them.

Halden Morris, senior lecturer responsible for technical and vocational education in the School of Education at the UWI, Mona (chapter 7), introduces a little researched area, quality assurance for tertiary TVET. In fact, Daniel, Alluri and Mallet (2008) refer to tertiary TVET as "still largely an untilled field". Normally, TVET, as presented by Paulette Dunn-Pierre in chapter 8, covers education that facilitates the acquisition of skills, knowledge and understanding necessary for employment in particular occupations; this is not usually considered to be the focus of tertiary education. Morris clarifies the concept of quality assurance for tertiary TVET and provides information on a successful quality assurance approach being employed internationally. He also presents indicators of quality for TVET at tertiary level and analyses these indicators in relation to what exists in the Caribbean context.

Dunn-Pierre, a former executive director of the Human, Employment and Resource Training Agency, Jamaica, argues for the importance of TVET in improving workforce competitiveness and redounding to the development of the nation and region. Chapter 8 presents a comprehensive discussion of the Caribbean Association of National Training Agencies (CANTA) standards for quality assurance in the TVET system. The Caribbean Association of National Training Agencies standards impact the delivery of training for and award of the Caribbean vocational qualification. She shows that CARICOM countries have been slow in taking up the Caribbean Association of National Training Agencies standards, however. The importance of TVET standards and quality assurance mechanisms in the region is reiterated.

Part 3, "Internal Issues", puts the focus on the individual higher educational institution and the myriad of issues that impact on quality assurance there.

There is an increasing recognition in higher education that "strategic planning can no longer be separated from quality assurance, nor can strategic planning or quality management work in a vacuum of information" (Shawyun, n.d.). The important nexus between quality assurance and strategic planning has often been recognized by the creation of one office to deal with both areas, as is the case at the Singapore Management University, which has an office of strategic planning and quality assurance. In other cases, like Northern Caribbean University in Jamaica, responsibility for quality assurance and improvement rests at the highest levels of the university – the president – the same office that is responsible for strategic direction (see Northern Caribbean University 2012). John Gedeon, senior planning officer, University Office of Planning and Development, in chapter 9, presents a basic guide to summative evaluation, exploring in detail all the components of the strategic management process using the balanced scorecard method. Gedeon makes the case for attention to summative evaluation at the end of a strategic planning period in light of the dearth of attention to this kind of evaluation.

The educational landscape in the Caribbean region is made complex by the presence of offshore providers, who claim that their franchised programmes are of the same quality as those on their home campuses. Similarly, Caribbean nationals also take advantage of cross-border educational offerings, usually by studying abroad. In chapter 10, Dameon Black of the University College of the Caribbean in Jamaica interrogates these claims. He argues that the tertiary level sector has become significantly more commercialized and therefore concerns with quality are even more pertinent. He then recommends a path that encourages creative engagement with international higher education providers in order to add value while assuring quality. In managing the process of international education, it is important that the extractive nature of the process be contained to ensure that the countries of the region benefit.

The complexity of the tertiary level education landscape in the Caribbean is reflected in the presence of cross-border education, as discussed by Black in chapter 10, online programmes in chapter 11 and local community colleges in various kinds of relationships with the UWI in chapter 13. Cobley (2000) notes that the presence of community colleges, which proliferated in the 1990s, has the potential to democratize access to higher education in the region. Cobley astutely discusses the historical process by which many of the community colleges in the Organization of Eastern Caribbean States were formed in a bid to provide access to higher education to students in the original "little 8" – the smaller territories in the Eastern Caribbean. S. Joel Warrican's contribution (chapter 13), on the process of creating the St Vincent and the Grenadines Community College, discusses this continuing dynamism in the sector in that part of the Caribbean. The formation of community colleges continues and these institutions have an important role to play in the national development agendas of various nations. Given the decreased possibility that the Organization of Eastern Caribbean States nations will be able to develop their own local universities any time soon, community colleges will be the logical solution. Warrican, former director of St Vincent and the Grenadines Community

College, gives an insider's view into the process of the formation of one such college, which provides useful and unique learning for similar activities in the region in the future. As Warrican indicates, questions of leadership, equity, access and resources need to be attended to in any process of amalgamation.

Online modalities are increasingly popular for delivery and reach of higher education. The population who favour and utilize online programmes is often different from the traditional student. Learning in the virtual space requires special and targeted forms of support. As chapter 11 makes clear, the question of the quality of online programmes is recurrent and resounding. Online programmes have a bad reputation in some circles. In the US context, this opprobrium is often directed at for-profit colleges, who are accused of "preying on students to rake in federal financial aid" (Kelderman 2011, B4).

The real question of quality in online learning is the question of quality in higher education. Quality is not just how many people graduate, but what those graduates know. Quality is also related to how long it takes, and how much it costs, to deliver that learning (Mendenhall 2011, B24).

Various mechanisms have been employed to evaluate the quality of online programmes. Pamela Dottin, programme officer of the QAU, UWI, with responsibility for the Open Campus, introduces one instrument that is available for the evaluation of online programmes, called the Shelton Scorecard (chapter 11). Dottin's analysis of this mechanism shows it to be a worthwhile instrument for such evaluations given its attention to transactional distance, for example, an area that is not covered by quality evaluation instruments for traditional face-to-face programmes.

Patrick Anglin, the officer currently in charge of policy and infrastructure in the Office of the University chief information officer, at the UWI, explores a cutting edge trend in higher education, which has not yet entered the higher education sphere in the Caribbean – the massive open online course (chapter 12). Very little appears to be known about this latest example of the use of technologies in education in the region and Anglin presents a comprehensive overview of the meaning and evolution of the platform as well as issues raised that many neophytes will find useful. He then turns the spotlight on what meaning massive open online courses could have for tertiary education in the region. In giving a number of alternatives for the higher education sector to approach massive open online courses, he asks that another look be taken at massive open online courses for the benefit of the region.

Using the UWI as a case study, and with UWI student feedback data as a backdrop, Sandra Gift, senior programme officer at UWI's QAU, with responsibility for the St Augustine campus, discusses selected quality standards and models, designed for the management of quality within organizations. Gift identifies those components that can contribute to the articulation of an eclectic quality model to move UWI up to the next level of quality in chapter 14. Using components of the selected International for Organization for Standardization (ISO) standards and quality models discussed, with elements of the ISO standard as the mainstay, she fashions an eclectic quality model for an integrated approach to academic and administrative quality at UWI – a QMS. Academic and administrative quality are the two bookends in Gift's QMS. This does not negate the continuing utility of the "fitness-for-purpose" approach that is applied to quality assurance reviews of academic programmes. Such an eclectic model can be used to guide planning and implementation of academic programmes and other activities and, along with the "fitness-for-purpose" approach, therefore, can ensure continuous improvement not only of teaching and learning but also of governance and management of quality and implementation of quality policy at UWI (and, by extension, other higher education institutions in the region). The focus on the ISO standard for developing and implementing a QMS is not new or singular. Other Caribbean higher education institutions like Northern Caribbean University and the Excelsior Community College in Jamaica are designing their systems around the ISO 9001-2008, "a globally accepted standard for providing assurance on the ability to satisfy quality requirements and to enhance customer satisfaction" (Northern Caribbean University 2012, 7). Gift's proposal, however, by its eclectic nature offers the benefits of all the approaches involved.

In practice, quality assurance entails numerous administrative transactions or details, as June Wheatley, a senior administrative assistant in the QAU of UWI highlights in chapter 15. Wheatley's is a unique voice in the field as she turns a spotlight unto the processes that are often overlooked or taken for granted in the big picture. The importance of efficient and effective administrative professionals in an organization, though obvious, is oftentimes undervalued. Indeed, the work of a quality assurance office depends significantly on trained administrators who take the lead on the detailed processes in a quality assurance review or evaluation. In the big picture approach to visioning quality assurance, the administrative transactions are taken for granted and often made invisible. Yet, they are an essential underpinning, as the quality assurance experience at UWI has shown. Wheatley's chapter explores quality assurance mechanisms from the perspective of the administrative support staff. It draws on experiences of such staff from the regional UWI, the UTech, Northern Caribbean University, and Excelsior Community College. In so doing, it re-establishes the importance of administrative support as significant partners in the quality assurance endeavour (indeed in higher education generally) and outlines ways of deepening the professional engagement/development of such staff.

Eduardo Ali, institutional effectiveness manager for UWI (St Augustine campus), explores the second bookend in Gift's pair of components in a QMS – service quality. He notes in chapter 16 how total quality management systems provide a management philosophy and framework for higher education improvement and customer service satisfaction. Ali maintains that many Caribbean universities and colleges have developed academic quality assurance systems but are now considering total quality management approaches with an emphasis on service quality. When service quality systems are focused on essential services, support leadership development and embrace organizational learning programmes, exceptional results can be achieved. In light of these points, Ali makes the case for total quality management, with an emphasis on new approaches in service quality measurement, as a transformational change agenda for higher education institutions in the Caribbean. The chapter refers to a regional research project that was performed by the author to understand service quality in Caribbean higher education, reflects on his past and recent experiences at UWI in developing service excellence initiatives and recommends capacity building of service quality practices and processes through leadership and employee learning.

The volume closes with an epilogue (chapter 17) that points a way forward for the conversation on quality assurance in higher education in the anglophone Caribbean. The dynamic nature of the quality assurance process ought to be reflective of the dynamism of higher education. Among the issues that are becoming increasingly relevant in quality in higher education in the Caribbean is the need for a recognized process for evaluating external quality assurance agencies, that is, "evaluating the evaluators", quality assuring MOOCs, and taking account of the development of global accreditation standards. Individual higher education institutions will be challenged to undertake institutional research to contribute to the development of models of quality assurance that can tackle continual problems such as grade inflation.

References

Beckles, Hilary, Anthony Perry and Peter Whiteley. 2002. *The Brain Train: Quality Higher Education and Caribbean Development*. Board for Undergraduate Studies.
Buckner, Elizabeth. 2013. "Opinion: Arab Universities Must Admit the Cost of Corruption". *Chronicle of Higher Education*, October 17. http://chronicle.com/article/Opinion-Arab-Universities/142317/?cid=gn&utm_source=gn&utm_medium=en.
CACET (Caribbean Accreditation Council for Engineering and Technology). N.d. General note. http://www.stluciaengineers.com/wp-content/uploads/2010/05/CACET-General-Note.pdf.
CARICOM (Caribbean Community) Secretariat. 2011. "Moves Afoot to Enhance Accreditation Systems in CARICOM Member States". Press release 343/2011, September 15.
Cobley, Alan G. 2000. "The Historical Development of Higher Education in the Anglophone Caribbean". In *Higher Education in the Caribbean: Past, Present and Future Directions*, edited by Glenford D. Howe, 1–23. Kingston: University of the West Indies Press.
Daniel, John, Krishna Alluri and Joshua Mallet. 2008. "Tertiary TVET: Pathways for Pioneers". Guest Address at the University of Vocational Technology (UNIVOTEC), Sri Lanka. Commonwealth of Learning, http://www.col.org/resources/speeches/2008presentations/Pages/2008-10-01.aspx?print=true.

El-Khawas, Elaine, Robin DePietro-Jurand and Lauritz Holm-Nielsen. 1998. "Quality Assurance in Higher Education: Recent Progress; Challenges Ahead". October. Graduate School of Education, University of California, Los Angeles.

European University Association. 2001. "Quality Assurance in Higher Education: A Policy paper of the European University Association", September 27. EUA Council, Dubrovnik.

Johnson, Jovan. 2013. "Students' Loan Bureau still Suffering from High Delinquency Rate". *Jamaica Gleaner*, October 18. http://jamaica-gleaner.com/latest/article.php?id=48733.

Kelderman, Eric. 2011. "Online Programs Face New Demands from Accreditors". *Online Learning: The Chronicle of Higher Education*, November 11, B4–B5.

Knapper, Christopher. 2006. "How the Way We Teach Affects Student Learning". *UWI Quality Education Forum* No. 12, January, 80–91.

Leo-Rhynie, Elsa, and Marlene Hamilton. 2008. "Creating a Culture of Quality: The Role of the University of the West Indies in Caribbean Education". In *The Caribbean Community in Transition*, edited by Kenneth Hall and Myrtle Chuck-A-Sang, 304–15. Kingston: Ian Randle.

Mendenhall, Robert W. 2011. "How Technology Can Improve Online Learning – and Learning in General". *Online Learning: The Chronicle of Higher Education*, November 11, B23–24.

Northern Caribbean University. 2012. "Northern Caribbean University Quality Policy", November 26. Quality Management and Assessment Department.

Sangster, Alfred. 1994. "Education and Training: Key Elements in the Development Process". In *Jamaica Preparing for the Twenty-First Century*, edited by Patsy Lewis, 204–10. Kingston: Ian Randle.

Shawyun, Teay. N.d. *Strategic Planning as an essential for Quality Assurance*. https://www.academia.edu/2148043/Strategic_Planning_as_an_essential_for_Quality_Assurance.

UWI (University of the West Indies). 2000. "The UWI Quality Strategy: The Quality Assurance System at the University of the West Indies". Office of the Board for Undergraduate Studies.

———. 2000/2001. "Quality Assurance Strategy: The System in Action". Office of the Board for Undergraduate Studies.

———. 2006. "Report of the Chancellor's Task Force on Governance of the UWI".

Whiteley, Peter. 1998. "Quality Issues in Caribbean Tertiary Education". Paper prepared for the 1998 Annual Conference of the Society for Research in Higher Education, University of Lancaster, UK. December 15–17.

———. 2002. "Quality Assurance in Selected Caribbean Universities". In *Adult Education in Caribbean Universities* edited by Ian Austin and Christine Marrett, 249–71. Kingston: UNESCO.

1 | Foundations

1 | Improving Learning through Effective Teaching

Anna-May Edwards-Henry

Effective teaching commences with the conviction that the learning that emanates from teaching is the only valid evidence of effectiveness (Cross 2003; Bain 2004). This conception of teaching effectiveness speaks to teaching as being transformative, and the quality of teaching is marked by the level of change exhibited in learners. To establish and maintain the quality of teaching and learning requires more than the intuitive practice of teaching and more than simply getting students through examinations. Biggs (2001) proposed the concept of transformation as the "focus of quality assurance mechanisms for maintaining and enhancing the quality of teaching and learning in the (higher education) institution" (221). Transformative teaching is undergirded by a number of ideas and ideals that include an understanding of the learning process as well as the theories that underpin this understanding. There is also the consideration that to engage in effective, high-quality practice we must be cognizant of the variety of factors that directly impact the learning process, as well as student motivations and methods of student engagement, that are all part of the learning landscape. This chapter reviews some of the fundamental concepts that would be minimally adequate to assist teachers in becoming effective teachers in higher education. The chapter speaks directly to preparatory knowledge that informs teaching for learning. It deals primarily with what I have identified as the six factors for effectively transforming learners into the desired graduates, as well as considers how teaching can be effectively implemented even within a context of linearity, as is the semester-based system. An underlying assumption to the conceptions espoused is that teaching that is focused on student learning is achieved through "a reflective and inquiring approach" (Ramsden 2003, 8). (Note: Throughout this chapter the word *teacher* is used as opposed to lecturer or instructor or any other term of similar connotation to emphasize the role of teaching.)

Six Factors of Effective Teaching

Content or subject matter is but one of *six* factors that must be taken into account before one can effectively engage the teaching/learning process. These factors I have defined as: the *learner*, the *learning process*, the *subject matter*, a *facilitating learning environment*, *time constraints* and the *teacher*. We will examine each of these factors in turn, highlighting how they contribute to the preparation for effectively engaging in teaching that transforms learners and, consequently, results in enhanced high-quality learning.

The Learner

The *learner* is here used collectively to refer to *all* students in the diverse forms in which they appear in higher education classrooms. Effective teaching begins with an understanding of learners as individuals, students and future leaders, who have their own understandings of their roles within the higher education system, which might not be consistent with one's teaching intentions and purposes. This is complicated by their varying levels of preparedness, learning styles, motivations and background, all of which influence how they approach learning. However, since higher education institutions universally describe their graduates as being self-reliant, critical thinkers, problem-solvers with world views that reflect global citizenry, who are able to move countries ahead in a competitive environment, quality teaching and learning must be able to transform these diverse learners into graduates that exemplify the identified characteristics. In short, the outcomes of higher education are "products" that have been fundamentally transformed into individuals that can command, control and effectively manipulate and utilize an unknown future. The main aspects of learners to be explored in this transformative process are their learning styles, the range of competences underpinned by their multiple intelligences and their motivations and preparedness for learning. The transformation commences with determining where students are in the learning process and providing them with the tools to become learners who can take charge of their learning and harness and hone their own energies to become knowledgeable and competent in chosen disciplines.

Learning Styles: A learning style is a student's consistent way of responding to and using stimuli in the context of learning. Learning styles are not really concerned with what learners learn, but rather how they prefer to learn. The works of several researchers, for example, Kolb (1984), Honey and Mumford (1982) and Fleming and Mills (1992), illustrate a variety of ways in which people prefer to engage learning. Individual learning styles can be determined by completing learning style inventories, which offer insight into how information may be presented to appeal to learners. Thus having students complete learning style inventories is a common strategy or occurrence among the tools to facilitate the learning process.

Multiple Intelligences: Quite apart from differences in learning styles, students have different capabilities and varied potential because of their differing intelligence strengths. For example, while some students are verbally able (linguistic, after Howard Gardner 1993), others are mathematically inclined (logico-mathematical) or have spatial intelligence that impacts learning in any particular course. Gardner also included among the intelligences he identified bodily kinaesthetic, musical, interpersonal, intrapersonal and naturalist. However, Gardner's intelligences may not adequately address the competencies of today's learners, who have particular intelligence in using technology that may be played to in the classroom or context of a course. So there may be a ninth intelligence that I will refer to as digital intelligence. Teachers may enhance student learning by making use of this evolving intelligence in preparing learning opportunities for today's students.

Student Motives and Motivations: In a learner-centred environment, it is important to build students' trust and positive perceptions by first finding out their motives and motivations for their presence in higher education classrooms. The dialogue must continue thereafter, and effective teachers build into their classroom experiences, opportunities for continuing discourse throughout course implementation. One way for a teacher to solidify knowledge about their students is to create learner profiles. These describe the characteristics of the learners that impact on their ability and readiness to learn, and include such items as reasons for studying, prior knowledge, prior study skills and study circumstances, in addition to personal characteristics.

Quality teaching thus requires a deep consideration of the identified characteristics of our learners to better meet their needs in the process of facilitating the learning process that transforms them into the desired graduates.

The Learning Process

More than a century of research on how people come to know shows that our knowledge of the learning process is still evolving and subject to theorizing, but there is sufficient evidence to produce a viable picture of the learning process. For example, it is now known that there are two major ways in which information gets into long-term memory where it has the probability of being constructed into knowledge.

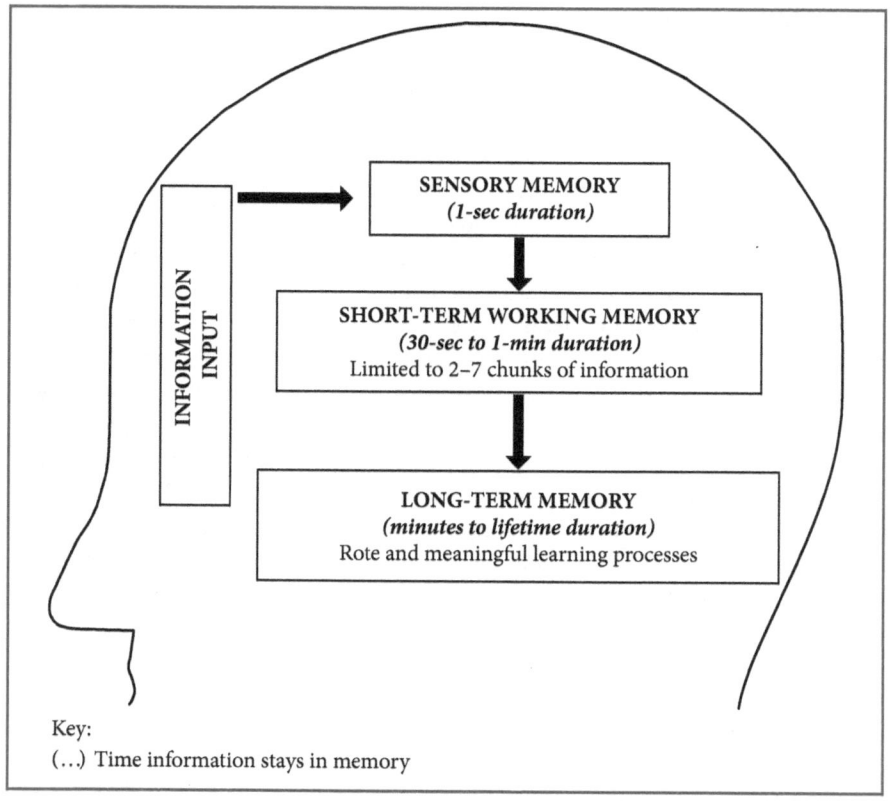

Figure 1.1. Processing information

These two ways of learning are (1) *rote* learning and (2) *meaningful* learning (Mayer 2002) (figure 1.1).

Rote learning is the typical means for acquiring first principles of which a wide range of Level I courses in disciplines offered at higher education institutions comprise. A similar approach is apparent when learners encounter material that does not make conceptual sense to them, and we may go right back to initial encounters with fundamental concepts, such as, the alphabet or multiplication tables. Rote learning involves repetition until the information gets into long-term memory. There are no prior frameworks to make meaningful links, so that for foundational and first principles, schema in the brain must be organized and developed from scratch, as it were. Such structures are generally weak, as retrieval can be problematic. Meaningful learning, on the other hand, is that by which learners make sense of the information as they interact with it (Ausubel 1960; Novak 2002). In attempting to learn meaningfully, new information entering the learners' brains meets old information and initially creates some dissonance. This information, which learners obtain from interacting with objects and events, allows learners to construct their own ideas about those objects and events. Learning comes about because learners keep fitting new information together with what is already known. This process essentially describes *constructivism*, a theory which von Glaserfeld (1989), has been credited with popularizing (Perkins 1991; Driscoll 2005). There are several implications of constructivism for teaching practice, especially in today's higher education classrooms. The constructivist classroom is organized to engage students by using active learning techniques, such as experiments and real-world problem solving, to create more knowledge. Students are then required to reflect on and *talk* about what they are doing and how their understanding is changing. However, one of the main challenges of constructivism is the fact that as learning is a personal and individual activity, it is quite possible for "mis-construction", where concepts are misconstrued and/or poorly or inappropriately formulated in the schema, or entirely misunderstood (Novak 2002). Unless the learners articulate or demonstrate in some way what they have constructed, their understandings are unknown, and there will be no opportunities to rectify their "mis-construction". In that way misconceptions may be permanently ingrained and then have a scaffolding effect on future learning. *Brain-based learning theory* offers support for the constructivist perspective in two very salient ways. The first is that there is a proven physical connection made in brain cells or neurons that relate to knowledge construction. These are the dendritic connections formed between them as neurons communicate during the learning process (figure 1.2). Learning takes place when neurons communicate. If information is incorrectly constructed, the physical structures, dendrites, are still created and it is only through extended periods of lack of use of these dendrites would they diminish. Secondly, several researchers (Greenough, Black and Wallace 1987; Diamond 1988; Healy and Harvey 1990; Kotulak 1996; Ramey and Ramey 1998) clearly show the development of more dendritic connections and heavier brains where there is an enhanced interaction with the environment, highlighting the enriched learning environment as facilitating improved or enhanced learning.

Quality teaching ensures that student engagement not only facilitates the development of dense neurons and neural networks by the interaction of learners with appropriate information, but must also ensure that mechanisms exist throughout the learning process to ensure that knowledge constructed is correct and correctly used.

The Subject Matter

The subject matter means knowledge of the content, borne out of a critical review and assessment of the discipline's concepts, constructs and epistemology. The subject

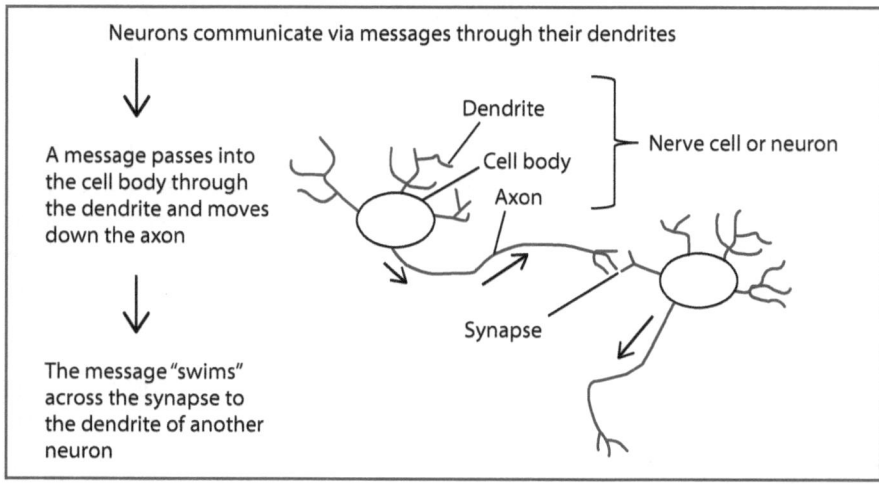

Figure 1.2. Nerve cells and communication

matter must be known from a meta-analytical perspective, which includes knowing about how knowledge of the subject is acquired and best understood, as well as "thinking about thinking" in the discipline. The effective teacher undertakes a critical analysis of the subject matter in terms of distinguishing foundational concepts that the students must know and broader concepts that are built on the foundational concepts; areas of the course that are conceptually easy for students to grasp, and more challenging areas (since the teacher will not treat easier and more difficult areas in the same way); areas that *must* be addressed in the earlier stages of student development in the discipline as opposed to areas that require application and analysis and other higher order thinking; and, areas that deal with procedural skills and knowledge. It is only when the course is conceptualized through this rigorous process can it be effectively broken up or modularized for effective interrogation by the students (Diamond 2008). This applies whether the course is mathematics, English, or engineering. Modularizing a course can be approached in a variety of ways, including following the chapters of a textbook, content/concepts of a course, time frame, a metaphor, and issues or combination of these or other constructs.

The significance of the conceptualization and modularization processes, however, is the promotion of opportunities for learners to build the right kind of schema in right ways leading to high-quality learning output.

A Facilitating Learning Environment

The learning environment is the context that the teacher provides to facilitate the learning process. The creation of the learning environment is the overt act of teaching.

Marzano et al. (1997) conceptualized learning as occurring in five dimensions that all contribute and interact to form a framework within which learning can eventually become a cognitive habit. The first of those dimensions is *attitudes and perceptions* and this is the dimension that relates to the type of learning environment teachers in higher education, in particular, are required to create. The other dimensions *acquiring and integrating knowledge, extending and refining knowledge, meaningful use of knowledge* and *productive habits of mind* unfold within the context of the dimension of attitudes and perceptions, which makes this dimension the critical one. Teachers ought to be able to create for students, environments that generate positive attitudes and perceptions evident as *classroom climate* (acceptable classroom behaviours by both teachers and peers as well as by a sense of order and comfort), and in the nature of *classroom tasks* (which should be interesting to students, and perceived by them as valuable and achievable). A major challenge to teachers is that today's students differ markedly from students in the past. It should therefore be a consideration that teaching methods which adequately served learners in former times may no longer be adequate to meet the needs and skills set of today's students or the competencies they are required to develop. Effective teachers, recognizing this, are open and willing to try different methods, and update and hone their teaching skills and techniques. In addition to the "normal" challenges posed by the diversity of students in any classroom, there is also the increasing size of higher education classes. Today's practitioners must make more concerted efforts in this context to know their students, because these classes naturally tend to be more impersonal, and students can become easily "lost", physically and psychologically. However, many teachers faced with increased class sizes use a reductionist approach to their practice. They resort to a lecture format with a focus on information dissemination, limited discourse with students and reduced opportunities for feedback from students. These strategies are the exact opposite of what should be done. The larger the class size the more it becomes important for teachers to strive to make the classes "appear" small by engaging in activities that give students a voice and opportunities for class participation and feedback (Brown and Race 2002). Knowing a range of teaching approaches and methods thus becomes an imperative.

Teaching Strategies and Methods: Table 1.1 describes five approaches or strategies for teaching and gives examples of relevant methods in each case. Especially in higher education classrooms, the default teaching method is that of the lecture. Undoubtedly, this method is the easiest to implement. However, it is often not the best method for student learning. In fact the work of several researchers (Stuart and Rutherford 1978; Erickson 1984; Astin 1985; Chickering and Gamson 1987; Knapper 1995) verifies the challenges to the lecture method in facilitating learning, and it has been conclusively proven by Astin (1985) and Chickering and Gamson (1987) that students have to be continuously active to sustain their concentration during the lecture and to facilitate subsequent retention. Thus, for the lecture to be an effective learning method, knowledge of how learning may be facilitated

Table 1.1 Teaching approaches or strategies

Teaching approach or strategy	Description
Direct instruction	Direct instruction is mainly teacher-centred and covers the most ubiquitous of methods. This strategy is effectively used for providing information or developing step-by-step skills. It also works well for introducing other teaching methods. Methods include: *mastery lecture, didactic questions, drill and practice, demonstrations, guides for reading, listening and viewing,* and *structured overview*
Indirect instruction	Indirect instruction is mainly learner-centred. In this approach, there is a high level of student involvement in observing, investigating, drawing inferences from data, or forming hypotheses. It takes advantage of students' interest and curiosity, often encouraging them to generate alternatives or solve problems. In indirect instruction, the role of the teacher shifts from lecturer to that of facilitator and resource person. The teacher arranges the learning environment and provides opportunity for student involvement. Methods include: *case studies, cloze tests, problem-based learning* and *graphic organizers* (e.g., *concept-mapping*)
Interactive instruction	This strategy covers a range of methods involving groups. Interactive instruction relies heavily on discussion and sharing among participants. Students learn from peers and teachers to develop social skills and abilities, to organize their thoughts and to develop rational arguments. Teacher skill is important in using interactive instruction methods in structuring and facilitating the development of the group. Methods include: *cooperative learning, discussion, brainstorming, interviewing, inquiry discussion* and *questioning*
Experiential learning	Experiential learning approaches are inductive, learner-centred and activity oriented. The emphasis in experiential learning is on the process of learning and not on the product. Personalized reflection about an experience and the formulation of plans to apply learning to other contexts are critical factors in effective experiential learning. Methods include: *field trips, simulations, role play, experiments, instructional games* and *synectics*
Independent study	Independent study refers to the range of instructional methods which are designed to develop individual student characteristics, including thinking, self-reliance and self-improvement. The study is usually directed or guided by a supervisor or teacher. Methods include: *assigned questions, programmed instruction* and *learning contracts*

during lecture sessions, and skills in appropriately structuring the lecture for student engagement and learning are necessary. In addition, andragogical principles specifically dealing with teaching adults (Knowles 1980) and perhaps even heutagogical principles (Blaschke 2012) relevant to teaching adults at a distance may also apply.

Time Constraints

When planning for teaching, practitioners need to recognize that the semester is *finite*. It is therefore necessary to select course content that can be reasonably managed by the students during the time available in a semester, which is usually less than the official semester weeks. This is the process of auditing a course. The auditing process takes place from the perspective of the students – based on the learning that the *average student* of the class can be reasonably expected to accomplish within the allotted time frame. Modularizing the course content (Diamond 2008) and constructing a course calendar are important strategies in support of the teaching/learning function. The methods identified for accomplishing learning tasks should also be considered, as some methods do take longer than others for the material to be fully internalized and understood by students.

The quality of both teaching and learning is compromised when due care and attention is not paid to the element of time and nature of tasks in facilitating learning. The face-to-face time between teacher and students is hardly adequate for learning efficacy, so that learning tasks outside the classroom must be complementary or additive and integrated into the developmental experiences of the learner and thus give a fillip to their learning.

The Teacher

The sixth factor to be considered in preparing teaching for improved learning is the teacher. To deal with the rapidly changing characteristics of students in the contemporary classroom, teachers should both know more about their students, and also about themselves as teachers. A good place to start in getting to know one's teaching self is by reflecting on what one actually does. The significance of engaging in the reflective process has been highlighted by Schon (1987) in his seminal work on professionalism. He referred to *reflection-in-action* and *reflection-on-action* as processes in which the excelling practitioner engages to develop and hone competence. This emphasizes that the practice of teaching, like any other professions, is best engaged through a deliberate and reflective process. From time to time, effective teachers reflect on their teaching philosophy, refine it and re-examine their practice, especially with regard to ensuring the quality of their activities in achieving desired outcomes. This is done with the view to improve on or sustain the effectiveness of their practice. Several researchers emphasize the interrelationships among teaching philosophies, conceptions about students as learners and practice (Chickering and Gamson 1987; Brookfield 1995; van Note Chism 1998). Brann et al. (2005) also highlight the relationship between teaching philosophy and specific teaching behaviours and modalities. All teachers in higher education are thus encouraged to subscribe to a teaching philosophy as a part of developing a reflective habit. To help in this regard, teachers can determine both their teaching styles and teaching perspectives which may be done using inventories that are readily available online.

Teaching Styles: Teaching styles refer to the preferred approach towards engaging learners. When teachers know about their teaching styles, they can better rationalize their actions with respect to providing opportunity for learning by their students. Grasha ([1996] 2002) recognized five teaching styles – *expert, formal authority, personal model, facilitator* and *delegator*. Each style bears special characteristics as well as advantages and disadvantages. The inventory developed by Grasha and Riechmann can be accessed at http://www.longleaf.net/teachingstyle.html.

Perspectives of Teaching: Not only are there teaching styles inventories to help teachers better come to know themselves, but there are also teaching perspectives according to Pratt (2007). Pratt based his perspectives on teaching on what he described as teaching BIASes (beliefs, intentions, assessment and strategies) that are based on the teaching dynamic involving the elements *learners, content*, and *teacher* and his or her *values*. He identified five perspectives or BIASes of teaching as follows: *transmission, apprenticeship, developmental, nurturing* and *social reform* perspectives based on the specific relationships among the elements. To find out one's teaching perspective, one may take the teaching perspective inventory, which can be accessed at http://www.teachingperspectives.com/drupal/.

The Teaching Context: A Context of Linearity

Fundamentally, there are many areas under the six factors that provide insight into how teachers in higher education may prepare themselves to improve student learning. Yet, there is still one major consideration that gives pause to how teaching can and should be engaged, and this is the teaching context itself. Teaching unfolds in a context of linearity, which is the semester system that is defined on a weekly basis. A course calendar clearly illustrates the linear way in which the semester unfolds and the context in which student learning opportunities are defined. However, learning hardly ever takes place in a linear fashion. As highlighted previously, learning takes place as a complex series of activities during which students build schema reflected in their understanding and utilization of the material, information and environment. The quality of the schema is determined by the quality and quantity of interactions, ways of interrogation and continuous reflection on the emanating or emerging learning. Thus, to implement teaching by disseminating information in a one-off fashion is not congruent with learning. There must be some consciousness in organizing courses for a semester to utilize activities that offset the linearity of the teaching setting. Teaching methods that tend to offset linearity are those that allow students to visualize course components, make links among concepts, allow for review and "recursiveness", for interrogating and reviewing course material, as well as relating course material to real-world experiences. Importantly, there are also mechanisms in such courses for questioning and feedback to project into future learning. For example, methods such as *concept-mapping*, *cooperative learning* or *case studies* illustrate these characteristics. Methods like these go beyond the linear dissemination of information to the kinds of processes that support learning while reducing the impact of the linearity of the semester system.

Conclusion

Quality teaching in higher education that is marked by continuous learning improvement is grounded in a philosophy that intricately links teaching with learning. To engage in effective practice, teachers should approach their practice from a systematic, reflective perspective involving the six factors identified for transforming students into desired graduates, and should take steps to develop and hone skills to make adjustments to teaching methods that facilitate the learning process. Teachers also have to contend with a system that is time bound and which unfolds linearly. This appears to be diametrically opposed to how learning occurs and speaks to the need for teaching methods to go beyond the one-off delivery of content. As a consequence of all of these factors that impact quality teaching, higher education teachers need to be fully equipped and prepared to meet the dynamism of the learning enterprise that is presented as today's classrooms. Their overt goal must be to attain high-quality teaching and sustained learning improvement towards the preparation of the desired graduates.

References

Astin, A. 1985. *Achieving Educational Excellence: A Critical Assessment of Priorities and Practices in Higher Education*. San Francisco: Jossey-Bass.

Ausubel, D.P. 1960. "The Use of Advance Organizers in the Learning and Retention of Meaningful Verbal Material". *Journal of Educational Psychology* 51 (5): 267–72.

Bain, K. 2004. *What the Best College Teachers Do*. Cambridge, MA: Harvard University Press.
Biggs, J. 2001. "The Reflective Institution: Assuring and Enhancing the Quality of Teaching and Learning". *Higher Education* 41 (3): 221–38.
Blaschke, L.M. 2012. "Heutagogy and Lifelong Learning: A Review of Heutagogical Practice and Self-Determined Learning". *International Review of Research in Open and Distance Learning* 13: 1.
Brann, M., C. Edwards and S.A. Myers. 2005. "Perceived Instructor Credibility and Teaching Philosophy". *Communication Research Reports* 22 (3): 217–26.
Brookfield, S.B. 1995. *Becoming a Critically Reflective Teacher*. San Francisco: Jossey-Bass.
Brown, S., and P. Race. 2002. *Lecturing: A Practical Guide*. London: Routledge.
Chickering, A.W., and Z.F. Gamson. 1987. "Seven Principles for Good Practice in Undergraduate Education". *American Association for Higher Education Bulletin*, March.
Cross, K.P. 2003. "Learning Is about Making Connections". UWI/Guardian Life Lecture, University of the West Indies, St Augustine, Trinidad.
Diamond, M. 1988. *Enriching Heredity*. New York: Macmillan.
Diamond, R.M. 2008. *Designing and Assessing Courses and Curricula: A Practical Guide*. 3rd ed. San Francisco: Jossey-Bass.
Driscoll, M.P. 2005. *Psychology of Learning for Instruction*. 3rd ed. Boston: Allyn and Bacon.
Erickson, F. 1984. "What Makes School Ethnography 'Ethnographic'?" *Anthropology and Education Quarterly* 15: 51–66.
Fleming, N.D., and C. Mills. 1992. "Not Another Inventory, Rather a Catalyst for Reflection". *To Improve the Academy* 11: 137–55.
Gardner, H. 1993. *Frames of Mind: The Theory of Multiple Intelligences*. New York: Basic Books.
Grasha, A. 1996 (2002). *Teaching with Style*. Revised, San Bernardino, CA: Alliance.
Greenough, W.T., J.E. Black and C.S. Wallace. 1987. "Experience and Brain Development". *Child Development* 58 (3): 539–59.
Healy, S.D., and P.H. Harvey. 1990. "Comparative Studies of the Brain and Its Components". *Netherlands Journal of Zoology* 40 (1–2): 203–14.
Honey, P., and A. Mumford. 1982. *Manual of Learning Styles*. London: P. Honey.
Kolb, D.A. 1984. *Experiential Learning: Experience as the Source of Learning and Development*. Englewood Cliffs, NJ: Prentice Hall.
Knapper, C.K. 1995. "Understanding Student Learning: Implications for Instructional Practice". In *Successful Faculty Development: Strategies to Improve University Teaching*, edited by W.A. Wright. Bolton, MA: Anker.
Knowles, M. 1980. *The Modern Practice of Adult Education: From Pedagogy to Andragogy*. Englewood Cliffs, NJ: Prentice Hall/Cambridge.
Kotulak, R. 1996. *Inside the Brain: Revolutionary Discoveries of How the Mind Works*. Kansas City, MO: Andrews and McMeel.
Marzano, R.J., D.J. Pickering, D.E. Arredondo, G.J. Blackburn, R.S. Brasat, C.A. Moffett, D.E. Paynter, J.E. Pollock and J.S. Whisler. 1997. *Dimensions of Learning*. Alexandria, VA: Association for Supervision and Curriculum Development.
Mayer, R.E. 2002. "Rote Versus Meaningful Learning". *Theory into Practice* 41 (4): 226–32.

Novak, J.D. 2002. "Meaningful Learning: The Essential Factor for Conceptual Change in Limited or Inappropriate Propositional Hierarchies Leading to Empowerment of Learners". *Science Education* 86 (4): 548–71.

Perkins, D.N. 1991. "What Constructivism Demands of the Learner". *Educational Technology* 31 (9): 19–21.

Pratt, D. 2007. "Can Excellence Take Different Forms in Scholarly Teaching? Exploring Our Teaching Biases". UWI/Guardian Life Premium Open Lecture, UWI, St Augustine, Trinidad.

Ramey, C.T., and S.L. Ramey. 1998. "Early Intervention and Early Experience". *American Psychologist* 53: 109–20.

Ramsden, P. 2003. *Learning to Teach in Higher Education*, 2nd ed. London: RoutledgeFalmer.

Schon, D. 1987. *Educating the Reflective Practitioner: Toward a New Design for Teaching and Learning in the Professions*. San Francisco: Jossey-Bass.

Stuart, J., and R.J. Rutherford. 1978. "Medical Student Concentration during Medical Lectures". *Lancet* 2: 514–16.

van Note Chism, N. 1998. "Developing a Philosophy of Teaching Statement". *Essays on Teaching Excellence: Towards the Best in the Academy* 9 (3): 1–2.

von Glaserfeld, E. 1989. "Constructivism in Education". In *The International Encyclopedia of Education, Supplement Vol. 1*, edited by T. Husen and T.N. Postlethwaite, 162–63. Oxford/New York: Pergamon.

2 | A Case for a Tertiary Education Council for the Caribbean Community

Halima-Sa'adia Kassim and E. Nigel Harris

Tertiary education is part of a larger public policy concern of nation-states to realize the goals of economic growth and consolidation, social equity and nation-building. Concern with optimizing the performance of the tertiary education sector continues to preoccupy individual tertiary level institutions, governments or both. The rising concern is a result of participation rates, the diversity of institutions, quality of programmes and the cost to the public purse. These concerns in turn have led to questions about the adequacy of existing institutional governance and management structures and processes to meet stakeholder expectations (Meek 2003). A structured and organized tertiary education system is seen as a mechanism to address the chaotic and unplanned development of the tertiary sector that lead to unnecessary duplication, uncertain quality control, inefficient linkages between education programmes and societal needs, variation in policies with respect to financing and insufficient attention to the roles various sectors of society must play in such financing.

Within the Caribbean Community (CARICOM), functional cooperation is seen as the pursuit of synergies and specific actions derived from combining resources for the purpose of realizing the benefits of economic development. The University of the West Indies (UWI) is arguably a good example of functional cooperation; UWI functions as one entity with common policies, common standards and a single governing body, the University Council, a system serving seventeen Caribbean anglophone countries which are member or associate states of CARICOM (Harris 2010).

A coherent and structured regional tertiary education system is another good example of functional cooperation. Knight (2012) suggests that a coherent and structured tertiary education system is also an example of "higher education regionalization" in which there is a process of intentionally building connections and relationships among higher education actors and systems in a region. Findlay and Tierney (2010), writing on tertiary education in the Asia-Pacific region, agree that "regional cooperation can add value by reaching a deeper understanding of the forces for change, sharing experiences to build confidence in the ability to adjust and to capture the benefits on offer, and removing impediments to integration" (1–2).

Examples of International Initiatives to Promote Coherence in Tertiary Education

Howe and Cassell (2002, 1) note that "worldwide, it has now become a compelling imperative that tertiary institutions engage in mutually beneficial dialogue, practical cooperation, and adopt strategies to effectively deal with the challenges of change, its attendant uncertainties, as well as position themselves to optimally exploit the manifold possibilities for new and exciting opportunities". The thrust towards cooperation through either project-driven approaches or through national education systems by way of tertiary education councils or commissions is seen as the mechanism to realize greater efficiency and coherence in the sector.

Tertiary education councils have been operating since the 1960s at the state or regional levels as in the United States, or at the national level in countries such as Botswana, South Africa, Rwanda, the Philippines, New Zealand and Mauritius. Councils are seen as a good way to more effectively and efficiently exploit opportunities and/or enable rationalization within the tertiary education landscape. Some councils may concern themselves with systems-level interfaces, including the relationship between the tertiary education sector and the lower levels of the education system. Others may focus on the specific sectoral relationships such as that between tertiary education institutions (TEIs) and the economy, while still others may address cross-border issues and global interfaces among the various tertiary education sectors and institutions. There is thus often great diversity in their vision, mandate, functions and objectives, as well as their legal standing.

Internationally, there has been an increasing convergence within the tertiary education system led by international or regional intergovernmental organizations. For instance, the efforts of the European Commission led to the creation of the European Higher Education and Research Area in which national higher education and research systems became more compatible and more comparable, thus facilitating increased interaction and mobility of students, graduates and staff across borders (European Commission 2011). The Pan-African University project formulated in 2011 – aimed at enhancing Africa's competitiveness and growth – established an academic network of already existing postgraduate and research institutions in East, West, Central, North and Southern Africa. It was seen as a major step in establishing the African higher education and research space initiated by the Commission of the African Union and the Association of African Universities. The Association of South East Asian Nations embarked on a programme to strengthen relations and activities among higher education institutions through the Association of South East Asian Nations University Network, while in Latin America and the Caribbean there is the Latin America and Caribbean Area for Higher Education, a regional platform formally created for the mobilization of projects and studies that support academic cooperation and knowledge sharing in the region (Knight 2012).

A proposal was made in 2006 for the creation of a Commonwealth tertiary education agency to provide advice and assistance to member states and national associations on policy, governance and management issues in tertiary education. The idea was later endorsed at the Eighteenth Conference of Commonwealth Education Ministers held in Mauritius in August 2012 (Commonwealth Secretariat 2012). The core activities of the Commonwealth Tertiary Education Facility, which is located in Malaysia at the National Higher Education Research Institute, University Sans Malaysia (USM), would be data collection, research and the production of policy papers. Also, included among the facility's activities are in-house projects related to an aspect of tertiary education in a particular region or member state and contract consultancies. The ministers also recommended the establishment of satellite units in other Commonwealth countries. Can the CARICOM region, with its fragmented system and diversity of actors look towards developing a similar system that promotes cooperation, convergence, collaboration, coherence and partnership within the tertiary education system?

The Call for a Coherent Tertiary Education System in the CARICOM Region

The emergence of tertiary level institutions in the Caribbean, its development and impact on the society, financing mechanisms, and governance and management of the sector have been well ventilated by Tewarie (2010), Ali (2007), Roberts (2003), Howe (2003), Peters (1993) and others. From the 1980s, the Caribbean tertiary education landscape responded to the liberalization and democratization of education which saw increasing demands for access and expansion in the number and type of tertiary institutions. The tertiary education sector in the Caribbean now encompasses a wide range of institutions that vary in type, size and mission.

The variety and diversity have created its own set of challenges that makes it imperative to address issues of duplication and redundancy, quality assurance, linkages between education programmes and societal needs, fragmentation, insufficient resources and inadequate cooperation that will enhance coordination, coherence and sustainability. To address these challenges, Ali (2007) and Tewarie (2010) recommended that a policy framework be developed and implemented to address governance, management, funding and regulatory issues. Further, the small size of the region and the commonalities of issues in the higher education sector make it prudent to pursue avenues for collaboration, partnerships and the creation of an integrated system rather than engaging in any excessive and unproductive competition for students, staff or financial resources (Howe 2003, 2011; Roberts 2009).

Edward Greene, former assistant secretary-general, CARICOM Secretariat, in a 2010 address to an OAS/UNESCO Conference on higher education in Suriname, highlighted the need for rationalization of programmes in higher education to complement rather than duplicate offerings and underscored the need for rationalization of the sector so that it is more efficient and relevant to the region. Sir Kenneth Hall, then pro-vice chancellor and principal of the UWI, Mona, argued for the creation of a tertiary education system which he saw as a network of institutions where "tertiary education institutions are linked to 'national and regional goals through an appropriate well-designed system characterized by a clear vision of the role of tertiary education, and willingness to respond to the changing societal environment in which it operates' " (2005, 27–28). This tertiary education system would make an important contribution to national and regional development. The case for a coherent and structured system then rests on the realization of economies of scale and scope.

Proposal for a CARICOM Tertiary Education System

The concepts of cooperation and coherence have infiltrated the reform proposals for the tertiary education sector. The organization of the sector was informally initiated by a TEI, namely UWI, to assist in the strengthening of these TEIs and to effect better articulation between their programme offerings and those of UWI. The purpose of this organization was to encourage increased cooperation and collaboration for the sector. The formation of the Association of Caribbean Tertiary Institutions, in 1990, filled a gap in regional cooperation and collaboration between TEIs in a wide range of academic and administrative areas such as standards, accreditation, and complementary development and delivery of particular programmes and services.

The vision for the tertiary education sector was also touted by the regional intergovernmental organization, CARICOM. The 1993 CARICOM Regional Education Policy focused on several fundamental areas of education and training from pre-primary to university, such as policy formulation, relevance, financing, management and administration, curricula development, access, training of teachers and other trainees, language and communication, standards of quality assurance and regional coherence. Carrington (1993), in detailing the need for regional coherence in the higher education sector, highlighted regionalism as an ideal, a resource and a style of operation. The 2001 CARICOM theme of "Investing in human resources with equity" emphasized, among other things, quality assurance, accreditation and rationalization of delivery as priority areas of focus. The 1993 and 2001 articulated higher education strategies demonstrated the concern with the creation and facilitation of more strategic links to realize the "added value" of collective action and cooperation and, thus, regionalism.

From about 2007, the region once again began to express a keen interest in strengthening the managerial leadership of the tertiary education sector. This was expressed through summits and declarations and, thus, took on more of a political tone. Prior to then, it was functional, focusing more on the practical aspects of the management of the sector. Miller, at the 2007 UNESCO conference in Trinidad, suggested that regional and functional cooperation in higher education and research will need to be a deliberate strategy for survival in the face of competition and globalization (Miller 2007, 16). He proposed the establishment of a coordinating and regulatory mechanism which would perform three functions: (1) promote exchanges on higher education policies, practices and programmes, (2) advise governments and institutions on higher education policies and (3) be empowered and mandated as the body to grant approval for the establishment and operation of higher education institutions across the subregion (17). The outcome document from the 2010 Caribbean Conference on Higher Education (UNESCO 2010) called on TEIs to actively pursue opportunities for cooperation with each other and to build on existing strengths in regional networks and with regional and international organizations and institutions to support academic cooperation and exchanges (UNESCO 2010).

Model for the Proposed Tertiary Education Council

Building on the earlier recommendations and actions from 1993 to 2010, UWI proposed to the Twentieth Meeting of the Council for Human and Social Development (COHSOD) held in October 2010 in Guyana:

> The construction of a regional tertiary education system, matched to some degree by similar national systems in countries with large and diverse tertiary sectors, could enable a more structured enunciation of the expectations of the tertiary education sector, provide policy guidance, ensure adherence to basic standards, enhance quality assurance, and provide general guidelines about financing that would enable its sustained growth and development consonant with global trends. The system – whether regional or national – would be overseen by a regional or national council made up of the various stakeholders of the sector – government, private sector, tertiary education representatives, civil society and students. (CARICOM Secretariat 2010)

The COHSOD "agreed to the establishment of the CARICOM Tertiary Education Council, which would have as its main remit, facilitating, within the Caribbean Community, the development and evolution of a harmonized system of TEIs and programmes, which complement and supplement each other, therefore reducing redundancies, duplication and waste" (CARICOM Secretariat 2011, 1). A proposal was later conceptualized on the idea of convergence where the focus was on the development of strategic links to realize economies of scale and scope while enabling TEIs to build up their capacities and capabilities to expand enrolment at a low cost with no additional expense to the government. The proposal was similar in nature to the previous idea of Howe (2003).

The proposal was refined with funding from the Commonwealth Secretariat and targeted stakeholder inputs from government, TEIs, professional organizations and intergovernmental organizations across the region. The tertiary education system was seen as one that would comprise a cluster of diverse tertiary institutions (for example, universities, community colleges, teacher colleges, specialized academies/institutions, private tertiary institutions and technical colleges), research institutions, regional policy setting institutions and other social institutions that influence the sector's development. The tertiary education system would have among its objectives:

- strengthening the system capacity and quality in all areas of the system including research
- strengthening the competencies, skills and knowledge required by all citizens
- improving alignment with national and regional development goals
- building/enhancing internationally competitive economies
- facilitating increased mutually beneficial linkages with stakeholders

Based upon work to support its objectives, the tertiary education system was seen as contributing to

- an increased percentage of the population with tertiary education qualifications
- improved quality and relevance of the education and training
- reduced disparities in access and equity
- more relevant and beneficial research in all areas of Caribbean society and economy
- improved sharing of data and greater transfer of knowledge, skills and technologies across the system
- increased number of graduates and citizens who meet the CARICOM and other national government profiles of the ideal Caribbean citizen

Each of the objectives was seen as directly and indirectly contributing to the realization of the combination of outputs. The tertiary education system should be guided by a core set of principles that would contribute to its efficient and effective functioning. These principles include student centeredness in terms of access, equity and service; excellence and quality; accountability, transparency and responsiveness of the institutions; institutions rooted in a Caribbean identity and producing the ideal Caribbean citizen (CARICOM Secretariat 1997); preservation of the uniqueness of each institution while maintaining diversity and institutional autonomy; and adequately financed with an element of its funding linked to performance.

Two models were considered in the creation of a structured tertiary education system supported by a governing body, a tertiary education council, both of which would be under the aegis of CARICOM and report to COHSOD. One model would be an umbrella body comprising, coordinating, advisory and regulatory functions with extensive responsibilities in areas such as quality assurance, accreditation, coordination and policy development and supported by a large bureaucracy. The second model would have a more contracted but functional role of coordination through policy development and advice for member governments. The latter model is pragmatic as it takes into consideration the area in which CARICOM has realized its most success – functional cooperation. The second model is also viewed as spurring the functioning of the CARICOM Single Market and Economy to promote a "more efficient operation of common services and activities for the benefit of its peoples" as identified in Article 8.i(i) of the Revised Treaty of Chaguaramas. Further, the establishment of this new entity was being proposed at a time when there was a mandate for the rationalization of regional institutions. As a result of global economic contractions, the region was economically challenged, and thus any entity which required a large injection of capital would not likely be well received by governments. The second model with its coordinating and advisory functions was seen as being more cost-effective.

The establishment of new organizations can be costly. For example, the CARICOM Secretariat, in 2006, had prepared a budget for the proposed regional accreditation agency of similar size and staffing based upon a 2007 salary scale and a ten per cent incremental increase on the preceding year. Based upon that calculation, salary emolument costs for the first year were projected at US$275,883 (not including first appointment and repatriation expenses) and for the second year were US$303,415. Operational costs for the first year were US$110,630 and for the second year were US$90,130. Using a system of consultancies to complete a two-year work programme for the tertiary education council, the CARICOM Secretariat in a draft project proposal estimated the total cost to be US$277,120. The Commonwealth Tertiary Education Facility is to be established at a minimal cost with the Malaysian Ministry of Education allocating one-off setting-up costs for the facility for a period of three years. After that, it is expected that the facility would be financially self-sustainable, that is, generate its own funds through sponsored research, partnerships, endowments, and so on (Commonwealth Secretariat 2014). It was estimated that the Government of Malaysia would contribute US$64,156.44 (RM200,000) to the initial cost of setting up of the Commonwealth Tertiary Education Facility and US$112,273.77 (RM350,000) to the annual operating budget. These budgets can be used as a benchmark in the initial setting up of the tertiary education council in the CARICOM region. Long-term funding for the council may also be realized through the establishment of a trust fund with contributions from international donors and sponsors, CARICOM member states, third states, and public and private entities. The funds could be held by the Caribbean Development Bank. To facilitate and drive development and innovation across the tertiary education sector and its representative professional organizations, a component of the trust fund should be devoted to supporting institutional development and innovation.

Three approaches may be considered with regard to staffing with implications for funding: (1) a small secretariat with staffing of five (executive director, two administrative officers and two technical officers and fixed term consultants, when required), (2) a small secretariat with a staffing of three (executive director, executive assistant and administrative assistant with fixed term contracts for professionals, when required, as is the Caribbean Accreditation Authority for Education in Medicine and other Health Professions model) and (3) the hiring of one technical person to be placed in the CARICOM Secretariat and who would have access to the resources of CARICOM Secretariat (the most cost-effective).

The approach to management of the tertiary education sector was a gradual, incremental non-binding one that allows critical mass to be achieved over time. It was a departure from the earlier recommendation of Miller (2007) and vision of Howe (2011) for a large body with extensive responsibilities. The model that was finally proposed was functional, in that the body would seek to identify policies or strategies to facilitate cooperation and convergence among national/regional TEIs, professional organizations, government and regional bodies responsible for policy-setting.

The proposal was taken to CARICOM at their May 2012 meeting of education and culture officials. Although the meeting accepted the idea for the establishment of a tertiary education council (sometimes referred to as the Regional Tertiary Education Council), in principle, concern lay with the operationalizing of it. In particular, concern lay with the extent of stakeholder consultation that had taken place around the proposal for a council and its financial sustainability. Officials doubted that the means were currently available to capitalize the proposed trust fund to the extent required. Accordingly, they resolved as follows: "*Agreed* and *advised* that the issue of approval for establishing a TEC [tertiary education council] would be deferred considering the need for much wider consultation with all the relevant stakeholders before considering the approval of establishing a TEC; *Noted* the need to be able to respond to logistical questions surrounding the establishment of a RTEC." The ministerial meeting that followed yielded similar results. They accepted the recommendation of the officials' meeting that there was a need for further stakeholder consultations with the tertiary education sector and requested clarification on the capitalization of the trust fund and the financial sustainability of the tertiary education council. The body requested this information prior to granting approval of the council at the next COHSOD.

The idea for some sort of agency or framework to facilitate coherence within the tertiary education sector is not new; it has been discussed for two decades (1993–2013). CARICOM, through the Standing Committee of Ministers Responsible for Education, agreed in 1997, and again in 1999, to the establishment of a regional accreditation mechanism as a means of facilitating the implementation of the free movement of skills within the CARICOM territories consistent with the provisions in the Revised Treaty of Chaguaramas. Given the delay in the implementation of the accreditation agency or other coordinating mechanisms, accreditation councils at the regional level to deal with the needs of particular disciplines such as medicine (Caribbean Accreditation Authority for Education in Medicine and other Health Professions), engineering (Caribbean Accreditation Council for Engineering and Technology) and regional professional tertiary education related bodies (for example, Association of Caribbean Tertiary Institutions; Association of Caribbean Higher Education Administrators; Caribbean Area Network for Quality Assurance in Tertiary Education) have been established to fill the gap operating within the region.

The tertiary education council as presented in 2013 at COHSOD did not find favour. This may be due to resource constraints of member states, relevance, given national managerial initiatives underway in the tertiary education sector, and a more pressing need to operationalize the regional accreditation body as a separate initiative, given the proliferation of higher education providers of dubious quality in the region. Even as CARICOM ministers placed on hold the creation of a structured and enunciated tertiary education system, the regional accreditation agency requires an additional signature for it to come into force, as was noted at the CARICOM Meeting of Education and Culture Officials in May 2013.

Alternative Responses

There are three possible responses to the leaner model proposed by UWI to CARICOM. One response is that no action is taken and the tertiary education sector continues as it is now with multiple professional tertiary education organizations, several proposed regulatory bodies and numerous tertiary education providers leading to even further uncoordinated and uneven development with no economies of scale and scope and high levels of inefficiency.

A second response is an informal lateral approach initiated by UWI to create an interuniversity council, which can facilitate dialogue and cooperation among the universities and colleges in the CARICOM region, as well as contribute to assisting in sustaining and improving the quality of public higher education. However, the voluntary nature of such an organization means that there is a high risk that successful implementation and outputs may be compromised by competing interests and lack of resources. Further, the universities may also compete against each other for the same resources and the market share.

A third response might be to seek support from Association of Caribbean Tertiary Institutions and Association of Caribbean Universities and Research Institutions using a bottom-up advocacy strategy to eventually secure buy-in for the idea of the council from the political stakeholders (government). This idea is similar to that posited by Hall (2005), who suggested building on the idea in the *UWI Strategic Plan 1997–2002*. The UWI would become the hub for this system working through Association of Caribbean Tertiary Institutions to support the change process and "would do all things necessary and feasible to enable tertiary level institutions to operate as effective functioning links to a seamless education system in the region" (UWI 1997 15).

Is This New Mechanism Feasible?

The tertiary education system, regardless of form, is only feasible insofar that there is institutional buy-in from member states, tertiary level institutions and related professional bodies. Equally important to its implementation is the allocation of core financial resources, if the expected outcomes are to be realized. Given the financial constraints of the region, it is doubtful whether it can support another institution or mechanism thus delaying the creation of a tertiary education council. Perhaps, more importantly, is that for either tertiary education council or the accreditation body to be a reality requires political will and the relinquishing of some level of independent sovereignty. Unless these conditions are realized, and at this point it is quite unlikely, the tertiary education council would not be a reality as conceptualized as an entity within the CARICOM.

A structured and carefully articulated tertiary education system supported by a tertiary education council will be operating in a vibrant knowledge and information society and it would need to respond quickly to changes. The extent to which the tertiary education council is embedded in or accountable to an intergovernmental organization with a bureaucratic structure could inhibit or delay decision making or pursuing action. Simultaneously, it must not be ignored that embedding or anchoring the tertiary education council to an intergovernmental organization like CARICOM can give it political and economic legitimacy. Based upon the above, given the hesitation by member states to commit to the accreditation agency, it is unlikely the tertiary education council would be a reality by 2015.

Conclusion

There is an urgent need for improved coordination for long-term planning and overall development of the tertiary education system. Given the heterogeneity within the tertiary education sector in the CARICOM region, there is a need for a structured and systematic approach that starts with the creation of a web of institutions held together by formal and informal rules and an oversight body that offers policy guidance, leading to the realization of economies of scale and scope. A tertiary education council is seen as the best mechanism for providing the enabling environment for the optimum functioning of the collective tertiary education sector. The overarching aim of such a process is to create a framework based on cooperation, functionality and exchange that is attractive to creating the ideal Caribbean citizen as well as building and enhancing work in areas of social relevance to support the optimum functioning of the CARICOM Single Market and Economy.

In looking ahead at the post-2015 education agenda, UNESCO notes that education system strengthening is a priority for effective governance, functioning and the assurance of a good quality education (UNESCO and UNICEF 2013). It is clear that the region has a short window of opportunity to reconsider the architecture and logistics of operationalizing the tertiary education council. A tertiary education council is critical to strengthening the capacity and quality of tertiary education so as to effectively contribute to national and regional development, competitiveness and innovation.

Postscript

Recently, an inaugural workshop and a pre-launch of the Commonwealth Tertiary Education Facility was held in November 2014 in Malaysia. Also, as part of the efforts to enhance cooperation and collaboration, at the end of the Eleventh Annual CANQATE Conference held in Belize in November 2014, a memorandum of understanding was signed by external quality assurance agencies that seek to promote free movement in the Caribbean to develop systems for mutual recognition, to share information and to harmonize aspects of policies in the development of best practices and to develop capacity building.

References

Ali, Eduardo. 2007. "Prospects for Enhancing Caribbean Higher Education Policy Research: The Trinidad and Tobago Model for Strategic Development of the Higher/Tertiary Education Sector". Paper presented at the UNESCO Forum on Higher Education, Research and Knowledge Second Regional Research Seminar for Latin America and the Caribbean. Trinidad, July.

CARICOM Secretariat. 1997. "Towards Creative and Productive Citizens for the Twenty-first Century and Human Resource Development and Science and Technology within the Context of the Single Market and Economy". Paper presented at the Eighteenth Meeting of the Conference of the CARICOM Heads of Government. Montego Bay, Jamaica. http://www.caricom.org/jsp/communications/meetings_statements/citizens_21_century.jsp?menu=communications.

———. 2006. "Working Document for the Fifteenth Meeting of the Council for Human and Social Development (COHSOD)". COHSOD 2006/15/2. Georgetown, Guyana, October.

———. 2010. "Draft Summary and Conclusions of the Twentieth Meeting of the Council for Human and Social Development (COHSOD): Education". Georgetown, Guyana.

———. 2011. "Proposal Outline to Commonwealth Secretariat: Support for the Establishment of a CARICOM Tertiary Education Council: Preparatory Activities of the Task Force". Georgetown, Guyana.

Carrington, Edwin W. 1993. "The Future of Education in the Caribbean: Report of the CARICOM Advisory Task Force on Education". CARICOM Secretariat, Georgetown, Guyana.

Commonwealth Secretariat. 2012. 18th Conference of Commonwealth Education Ministers (18CEM) Mauritius Communiqué, August. http://www.thecommonwealth.org/document/181889/34293/35232/249627/18ccemcommunique.htm.

———. 2014. "Development of a Business Plan for Commonwealth Tertiary Education Facility (CTEF)". http://thecommonwealth.org/assignment-development-business-plan-commonwealth-tertiary-education-facility-ctef#sthash.epIAic5Q.dpuf.

European Commission. 2011. "Commission Staff Working Document on Recent Developments in European Higher Education Systems Accompanying the Document Communication from the Commission to the European Parliament, the Council, the European Economic and Social Committee and the Committee of the Regions Supporting Growth and Jobs: An Agenda for the Modernisation of Europe's Higher Education Systems". SEC 1063, Brussels.

Findlay, Christopher, and William Tierney. 2010. "Introduction and Overview". In *Globalisation and Tertiary Education in the Asia-Pacific: The Changing Nature of a Dynamic Market*, edited by Christopher Findlay and William Tierney, 1–16. Singapore: World Scientific.

Hall, Kenneth O. 2005. "Developing a National Tertiary Education System, and the Changing Role of the University of the West Indies". In *Revisiting Tertiary Education Policy in Jamaica Towards Personal Gain or Public Good?*, edited by Rheima Holding and Olivene Burke. Kingston: Ian Randle.

Harris, E. Nigel 2010. "Proposal for the Establishment of a Regional Tertiary Education Council". Paper presented to the Twentieth Meeting of Council for Human and Social Development (COHSOD), Georgetown, Guyana. October.

Howe, Glenford. 2003. *Contending with Change: Reviewing Tertiary Education in the English-Speaking Caribbean*. Caracas: International Institute for Higher Education in Latin America and the Caribbean.

Howe, Glenford. 2011. "Transforming Tertiary Education in the Caribbean: Imperatives for a Regional Tertiary Education System (RTES) and Council (RTEC)". Paper prepared for the CARICOM Secretariat, Georgetown, Guyana.

Howe, Glenford, and Daphne Cassell. 2002. "A Rationale for Reconfiguring Tertiary Education in Montserrat & the CECS to Meet the Life-long Learning Challenges of the Twenty-first Century". Paper presented at the First Montserrat Country Conference, Montserrat. November. http://cavehill.uwi.edu/bnccde/montserrat/conference/papers/filename.html.

Knight, Jane. 2012. "Conceptual Framework for the Regionalization of Higher Education: Application to Asia". In *Higher Education Regionalization in Asia Pacific: Implications for Governance, Citizenship, and University Transformation*, edited by John N. Hawkins, Ka Ho Mok, and Deane E. Neubauer, 17–36. New York: Palgrave Macmillan.

Meek, Lynn V. 2003. "Governance and Management of Australian Higher Education: Enemies within and without". In *The Higher Education Managerial Revolution?*, edited by Alberto Amara and Lynn V. Meek, 179–201. The Netherlands Springer. doi:10.1007/978-94-010-0072-7_9.

Miller, Errol. 2007. "Research and Higher Education Policies for Transforming Societies: Perspectives from the Anglophone Caribbean". In *Selected Proceedings of the 2nd Regional Research Seminar for Latin America and the Caribbean: UNESCO Forum on Higher Education, Research and Knowledge Research and Higher Education Policies for Transforming Societies: Perspectives from Latin America and the Caribbean.* Paris: UNESCO.

Peters, Bevis. 1993. *Emergence of Community, State and National Colleges in the OECS Member Countries: An Institutional Analysis*. Cave Hill, Barbados: Institute of Social and Economic Research, UWI.

Roberts, Vivienne. 2003. *The Shaping of Tertiary Education in the Anglophone Caribbean Forces, Forms and Functions*. London: Commonwealth Secretariat.

———. 2009. "Balancing Quality and Quantity in Education: The Caribbean Challenge". Paper presented at the Sixth Annual CANQATE Conference, Barbados, October.

Tewarie, Bhoendradatt. 2010. "Concept Paper for the Development of a CARICOM Strategic Plan for Tertiary Education Services in the CARICOM Single Market and Economy (CSME)". Arthur Lok Jack Graduate School of Business, University of the West Indies, St Augustine, Trinidad and Tobago.

UNESCO. 2010. "Declaration of Paramaribo First Caribbean Conference on Higher Education: Integration and Development of the Caribbean". International Institute for Higher Education in Latin America and the Caribbean and Organisation of American States, Paramaribo, Suriname. April. http://www.unesco.org.ve/dmdocuments/declaration_caribbeanconference2010.pdf.

UNESCO and UNICEF. 2013. "Education in the Post-2015 Development Agenda Draft Synthesis Report of the Global Thematic Consultation on Education". May 2013. http://www.worldwewant2015.org/node/327364.

UWI (University of the West Indies). 1997. *The University of the West Indies Strategic Plan, 1997–2002*. Kingston: University of the West Indies.

3 | The Quality/Financing Conundrum in Caribbean Higher Education

Kofi K. Nkrumah-Young

Regardless of the mode for financing higher education, be it state, private or mixed, the question of the role of quality assurance often arises. The following questions are frequently asked:

- Should the state fund poor quality higher education institutions (HEIs)?
- Should the state channel its funds to those institutions that are considered to be delivering good quality and ignore those in need to improve their standards?
- Should funding be used as an incentive to improve quality?

In considering the link between quality and funding, this chapter discusses the main features of the quality assurance mechanisms existing in the Commonwealth Caribbean higher/tertiary education systems. It also outlines the resource allocation mechanisms being used in the region, and assesses the quality/resource allocation link in the systems, using the Massy (1996) framework and the Orr (2005) trajectory.

Methodology

The research for this chapter was undertaken using qualitative methodologies with data gathered primarily from source documents such as publications, minutes of meetings and correspondences as well as interviews. Analyses were done using inductive logic (Creswell 2003). The focus of the research was on the anglophone Caribbean, which comprises the fourteen English-speaking full member states of Caribbean Community. Purposive sampling was used to identify Jamaica, Trinidad and Tobago, and Barbados as representative for the region. Forty-five per cent of the anglophone citizenry reside in these countries; the region's largest tertiary education system is found there also. The tertiary education system in these countries comprises the regional university, the University of the West Indies (UWI), as well as national tertiary institutions. In Jamaica there are, in addition to the UWI Mona campus, the University of Technology, Jamaica (UTech), the Northern Caribbean University, the International University of the Caribbean, several teachers' training colleges, other colleges providing specialized training such as the Mico University College, and the community colleges. Barbados has, in addition to the UWI Cave Hill campus, the Barbados Community College, the Erdiston Teachers College, the Samuel Jackman Prescod Polytechnic, Barbados Institute for Management and Productivity and the central office of the UWI Open Campus. The tertiary system in Trinidad and Tobago consists of three publicly funded universities and colleges (including the UWI), training and technical vocational institutions and a range of administrative arrangements with foreign professional bodies, technical and vocational institutions, colleges and universities largely with private tertiary education providers (http://www.unesco.org/new/en/education/resources/unesco-portal-to-recognized-higher-education-institutions/dynamic-single-view/news/trinidad_and_tobago-1/). All three islands are also served by the UWI Open Campus, offering distance and online, as well as face-to-face education.

The Quality-Resource Allocation Link

Harvey and Knight (1996) outline five views of quality; two of these are relevant to this discussion – "the exceptional" view judges quality on the performance of the input factors and the "value for money" view relies on the end product for its judgement. They opine that the quality of inputs do not necessarily result in the same quality of outputs because there may be problems in the processing. There is, however, no uncertainty about ex post judgement of a finished product. The "value for money" view therefore provides a better basis for judging quality. This supports Layzell's (1998) view that inputs resource allocation models ignore quality as they focus on the "cost to continue". On the other hand, performance resource allocation models, he posits, were merit-based and, by rewarding improvement in outputs, cause more to be done with less.

There is also the argument about the "quality trap", which claims that to allocate resources to an institution for good performance and punish another for failure would result in the good becoming better and the bad becoming worse. Sir Howard Newby, then chief executive of the Higher Education Funding Council of England (HEFCE), reiterated this view in an address to the first cohort of students of the DBA in Higher Education Management at the University of Bath (February 17, 2003). In answer to a question, Newby pointed out that, unlike the allocation process for research, the Higher Education Funding Council of England limited the use of quality factors for allocating resources for teaching because it would result in punishing the students who were not at fault. He further stated that in the case of research, it was only the institution that suffered as a result of poor assessment. This view indicated a preference for an input resource allocation model because it seemed to stress the point that in order to improve quality, funding should be provided based on the input factors of the educational production process. A similar view was detected in the British government's white paper, *The Future of Higher Education* (2003), which proposed a rebalancing of funding based not only on research and student numbers but on strength in teaching. The document also proposed individual rewards for teaching excellence.

Burke (2002) points out that the difficulty in linking quality to performance rests with the inability of the performance indicators to "capture fully the essential but elusive character of quality in higher education" (45). Massy (1996) believes, however, that one cannot simultaneously pursue high-assay investment, avert the consequences of poor performance, and contain unit costs because something must give and that something depends on the expected academic unit's behaviour and the restructuring environment within which the resource is embedded. The effects of performance funding are captured in table 3.1.

The Massy framework suggests that

1. In a heavily centralized environment, where the institution or academic unit had no discretion about quality, it did not matter what resource allocation model was used, as incentives did not matter.
2. Where there is no link to performance, there is no incentive to either maintain or improve quality.
3. Linking funding to performance provides incentives to maintain and improve quality
4. Offering reward and penalties, however, would result in the quality trap in cases where there were insufficient insights about quality and the requisite services needed to support the growth of quality consciousness.

Massy (1996) further suggests that, in the case of rewards and penalties, an effective quality resource allocation model alongside the symmetric performance-based funding could mitigate the problem of the "quality trap". Massy concludes that in order to assess the effects of any resource allocation model on quality it would be

Table 3.1. Effects of performance funding

Funding vs performance	No discretion (A) Q = Design Q	Departmental discretion on quality	
		(B) Q = Frontier	(C) Q < Frontier
No linkage with performance	Incentives do not matter Fails on investment criterion	No incentive for maintaining Q Fails on investment criterion	No incentive for maintaining Q Fails on investment criterion
Positive linkage with performance	Incentives do not matter Partly meets investment criterion	Provides incentives for exemplary Q Partly meets investment criterion	Provides incentives for exemplary Q Partly meets investment criterion
Symmetric linkage with performance: rewards and penalties	Incentives do not matter Fully meets investment criterion Cannot fall into the "quality trap"	Provides full incentives Fully meets investment criterion Cannot fall into the "quality trap"	Provides full incentives Fully meets investment criterion May fall into the "quality trap"

Source: Massy (1996, 320).

necessary to address two questions: (1) Are the HEIs operating at or below the quality frontier? (2) Does the resource allocation model support the HEI's discretion on quality?

Whichever answer is given to the above questions results in a particular position on the matrix and in turn suggests the effect of the funding on quality, that is, either failing or meeting the investment criterion or resulting in the quality trap.

Orr (2005) points to another link between the resource allocation model and a quality assurance mechanism. According to him, funding provides a steering function for the HEIs (pointing in a desired direction), while the quality assurance mechanism performs a mapping function (clarifying or explaining the performance of universities). In order to properly indicate the effects of the funding on the quality of the institution, Orr argues, that it is necessary for both these factors to complement each other. For that reason, he concludes that the type of quality assurance mechanism employed should positively correlate with the funding method. This is summarized in Orr's trajectories in figure 3.1.

The funding methods are plotted on the vertical axis, of figure 3.1, and the quality assurance mechanisms are on the horizontal axis. The resource allocation mechanism and the quality assurance mechanism are divided in three groups each. Discretionary budget, as a form of resource allocation, includes all funding methods

The Quality/Financing Conundrum in Caribbean Higher Education 53

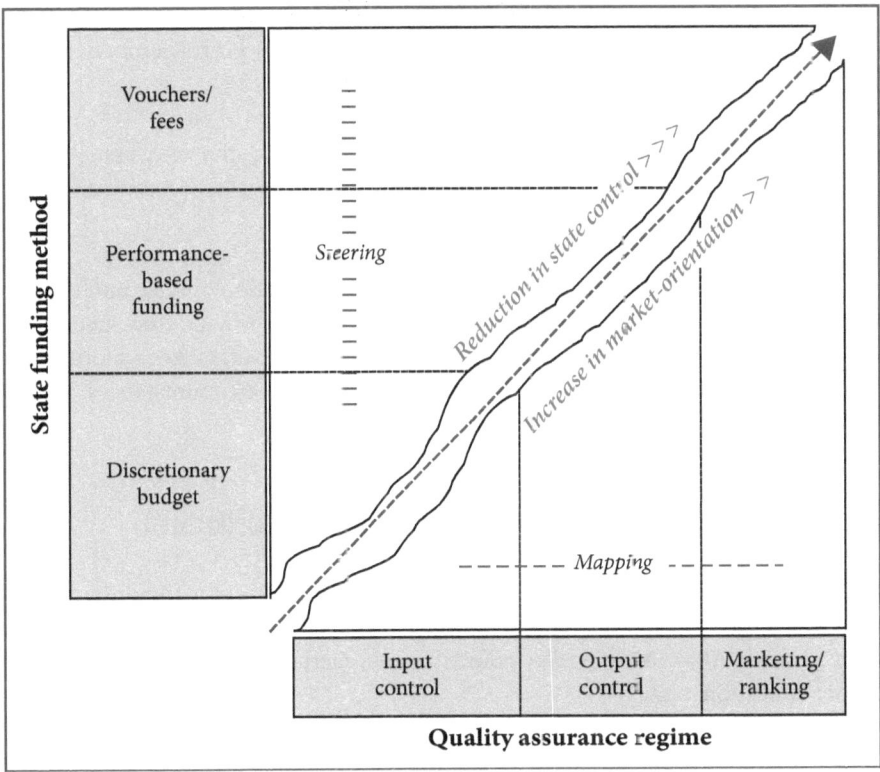

Figure 3.1. Orr's trajectory

whereby the state allocates resources by way of detailed input methods. The other methods are performance-based funding, and vouchers and fees. Input control quality assurance mechanisms correlate with discretionary budgeting as they judge the effects of the funding on the input mechanisms. Compulsory accreditation systems are classified as input control quality assurance mechanisms, which provide the state with input standards against which it judges the institutions. Such a system, Orr explains, could stifle innovation if the standards are too conservative. This has resulted in the move to more output control measures whereby the institution is responsible for producing a particular number of graduates, who are funded by the state. As such, the output quality assurance mechanism is congruent with performance funding. Voluntary accreditation systems are classified in the marketing category. This is because the provider may choose to participate in order to better position itself in the market. The output mechanism is deemed more suitable for a voucher system as it presents the opportunity for students to be informed about the "better institutions". Orr further uses two questions to apply his trajectory in assessing the funding and quality. These are:

- Does the funding method correlate with the method of quality assurance?
- Does the quality assurance method enhance the information provided via the funding method, or is it just bureaucracy?

The Orr trajectory suggests that the resource allocation model and quality assurance mechanism need to be correlated in order to provide information for determining the effects of funding on quality. Massy, however, makes a more direct link as he believes that performance funding could in, some circumstances, lead to improvements or maintenance of quality.

Quality Assurance Mechanisms and the Caribbean

Quality assurance is defined by Mishra (2006) as "the process of maintaining standards reliably and consistently by applying criteria of success in a course, programme or institution" (88). Mishra also points to four methods of assessing quality in higher education, namely:

- Self-evaluation – a process of continuous improvement.
- Best-practice benchmarking – "a tool used to improve products, services or management processes by analysing the best practices of other organizations to determine standards of performance and how to achieve them in order to increase customer satisfaction" (Lewis and Smith 1994, as quoted in Mishra 2006, 87).
- External quality monitoring – "an impartial and objective mechanism for assessing the educational institution by a peer team not directly related to the institution" (Mishra 2006, 33).
- Market-driven approach – the process of ranking institutions by media organizations such as the *Economist, Business Week, Financial Times* and *India Today*.

In reviewing the quality assurance procedures in higher education across the Caribbean, they are seen to take the form of EQM backed by self-evaluation through the work of national external quality assurance agencies. Best-practice benchmarking, which may be a part of the self-evaluation exercise, by some institutions, is not observed to play a major role. In Jamaica, the University Council of Jamaica (UCJ) registers institutions and accredits programmes of study at tertiary level institutions; it has the legal remit to award academic degrees, diplomas and certificates to individuals who have undertaken studies at prescribed institutions. It was established by the UCJ Act of 1987.

The Accreditation Council Act of Trinidad and Tobago Act (2004) empowers the Accreditation Council of Trinidad and Tobago (ACTT) to conduct and advise on "the accreditation and recognition of post-secondary and tertiary educational and training institutions, programmes and awards, whether local or foreign, and to promote quality and standards for post-secondary and tertiary education and training in Trinidad and Tobago". Any institution requiring any form of public funds from the Government of Trinidad and Tobago must be recognized either through registration or accreditation by ACTT. In addition, the particular programme must be approved by the ACTT.

The Barbados Accreditation Council is empowered by the 2004 Barbados Accreditation Council Act, *inter alia*, to register local, regional and foreign-based institutions that offer educational programmes in Barbados, to maintain a record of all institutions that are registered and accredited by the council and re-accredit programmes of study and institutions operating in Barbados.

All three external quality assurance bodies follow similar registration and accreditation procedures:

1. HEI provider applies to the council and then submits a self-evaluation report along with supporting documentation.
2. The council reviews the application and responds to the provider.
3. If the response is favourable, a date is set for a visit to the provider. A team composed of competent professionals is selected by the council to evaluate the programme/provider.
4. The provider is visited by the evaluation team.
5. The evaluation team makes an appraisal of the programme/provider and submits a written report to the council. The team report is sent to the provider, excluding the recommendation, for comments.
6. The team's and provider's reports are reviewed by the council.
7. The decision is communicated to the provider along with an evaluation report.

All the national external quality assurance bodies in the Caribbean are members of the Caribbean Area Network for Quality Assurance in Tertiary Education (CANQATE); there are plans afoot through the auspices of Caribbean Area Network for Quality Assurance in Tertiary Education for the formal mutual recognition of accreditation granted by each other.

It should be noted also that the Barbados Accreditation Council and Accreditation Council of Trinidad and Tobago Acts empower those bodies specifically to accredit institutions and programmes, but, unlike the UCJ, they do not have the powers to grant degrees. The UCJ Act does not mention institutional accreditation and, up to 2012, it had only accredited programmes and registered institutions. However, in 2012, for the first time, it accredited an institution, namely, the UWI, Mona. By institutional accreditation, the UCJ explained, it validated the quality standards of the UWI and recognized the internal quality processes for programme development, approval and review (interview with Grace Gordon, August 22, 2013). As part of the process to establish the Jamaica Tertiary Education Commission, there are discussions to revise the UCJ Act to reinforce its focus on the core functions of accreditation and quality assurance and to remove the degree granting authority (www.acticarib.org/docs/Legal_Side_of_Governance.pptx).

Apart from national recognition and programmatic and institutional accreditation, institutions in the Caribbean also seek and receive programme accreditation from regional and international bodies. These include, for example, the Caribbean Accreditation Authority for Education in Medicine and the other Health Professions, which accredits the Bachelor of Medicine and Bachelor of Surgery degree programmes of Jamaica, Barbados, and Trinidad and Tobago, and the other islands as well as the veterinary and dentistry schools in Trinidad and Tobago. An international body, the Association of MBAs, accredits the Mona and the Arthur Lok Jack Schools of Business in Jamaica and Trinidad and Tobago, respectively. Additionally, several programmes in the Faculty of Engineering, UWI, in Trinidad, are accredited by various international accreditation organizations such as the UK-based Institution of Chemical Engineers and the Royal Institute of Chartered Surveyors (http://sta.uwi.edu/eng/Accreditation.asp). At UTech, the Bachelor in Engineering (BEng Electrical Engineering and BEng Mechanical Engineering) is accredited by the Institution of Engineering and Technology and the Bachelor and Master of Arts in Architectural Studies are accredited by the Caribbean Association of Architects.

The above reflects quality assurance processes and mechanisms for teaching. There are no structured, stand-alone processes for assessing the quality of research being done in the Caribbean higher/tertiary education systems. Within the institutional accreditation process of the UCJ, however, Standard 2 states, "The institution systematically reviews its institutional research efforts, evaluation processes, and planning activities to determine effectiveness." In so doing, UCJ takes account of the research efforts but not in any sustained or structured fashion. Unlike in the United Kingdom, however, there is no research assessment exercise or similar system in the Caribbean.

Financing Higher Education in the Caribbean

The three Caribbean countries being examined operate a two-tier system involving both direct state support to the UWI and national higher/tertiary education

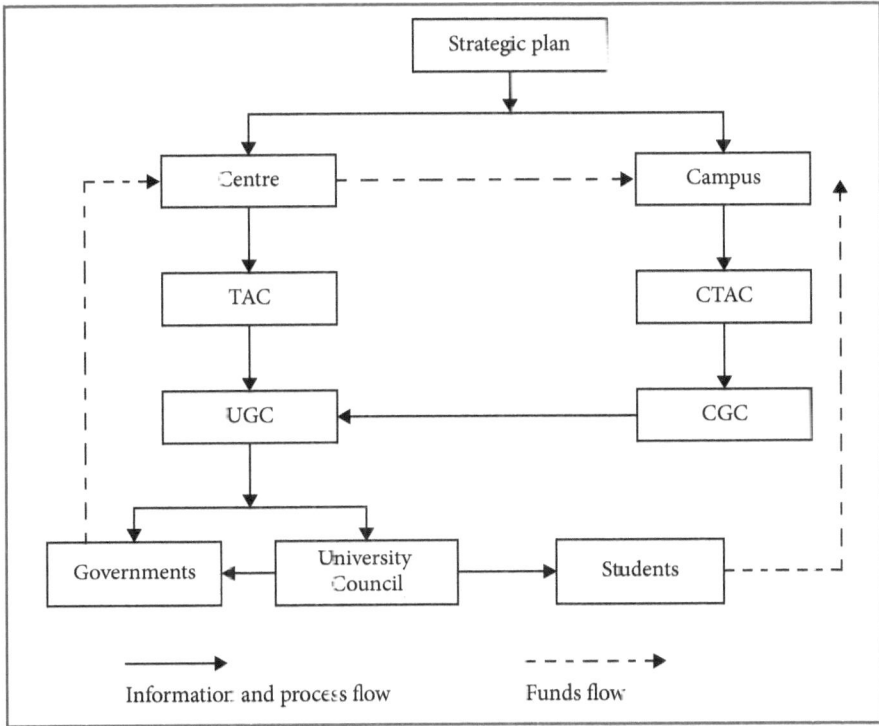

Figure 3.2. Resource allocation process for the UWI since 1994

institutions, as well as charging fees to students. The latter is handled differently in each country. In the case of the regional university, financial support from the state is determined by the process illustrated in figure 3.2.

As noted previously, the UWI operates campuses in Jamaica (Mona), Trinidad and Tobago (St Augustine) and Barbados (Cave Hill) as well as the Open Campus, which offers online, blended and face-to-face programmes. Co-ordination of the campuses takes place at the regional headquarters, called "the Centre". As shown in figure 3.2, the planning for the resource requirements for the Centre and the individual campuses is done separately; albeit to achieve the strategic direction of the University. The resource requirements for the Centre are first submitted to the Technical Advisory Committee for review. The Technical Advisory Committee then recommends the total budget of the Centre to the University Grants Committee for approval. The requirements for each campus are routed to their respective Campus Technical Advisory Committee for detailed review. The Campus Technical Advisory Committee thereafter recommends the campus budget to the Campus Grants Committee, which again considers and recommends acceptance to the University Grants Committee.

Funding for research at UWI is woven in the funding of teaching (Nkrumah-Young 2005) as funding from the government is mainly to cover salaries. Former minister of education for Jamaica Maxine Henry-Wilson attributed the salary disparity between UWI staff and the national tertiary level institutions to the research function; she stated that members of UWI staff are compensated for their research work (interview with Henry-Wilson, February 13, 2004). According to her, "To some extent, what we have now in terms of disparity relates to qualifications that were required and years of involvement in teaching, research and publication." This viewpoint, Nkrumah-Young (2005) points out, was supported by the bursar of the Mona campus, Elaine Robinson, who revealed the following:

> [W]e are funded as a research institution. It therefore means that essentially, we do not get any additional funds from the government specifically identified as research – our research is really built into our staff allocation in terms of work so an academic staff person is allocated a certain number of hours for teaching and research and is expected to allot a substantial amount of the summer period towards research so the salary structure really has built into it the research type and element in terms of cost. (Interview, November 18, 2003)

The contributing governments are expected to be bound by the decision of the University Grants Committee about the budgets; this was not always the case as examples of government shortfalls have had a significant impact on the financial state of the UWI. In cases where fees are charged to students, the council approves such fees.

The resource allocation process for the national HEIs in Caribbean Community is captured in figure 3.3. Generally, the institutions are required to make annual submissions to their respective ministries of education that review and recommend expenditure to the respective ministries of finance.

In the case of Jamaica, the government provides funds for the salaries of the national tertiary level institutions as per a predetermined pay level structure. The tertiary institutions are ranked in the following order: UWI, UTech and the colleges. Salaries for UWI staff are negotiated between the UWI council with the participation of the Ministry of Finance of Jamaica and the staff unions. The salaries for staff of UTech are negotiated between the Government of Jamaica and the staff unions of that institution. The government's team comprises representatives of the ministries of finance and education, and UTech. For the colleges, the negotiations are between the government (that is, the ministries of finance and education) and the Jamaica Teachers Association or the Bursars Association of Jamaica. Observation of the process revealed that UWI salary levels were used as the upper limit, UTech's were determined at the second level and the colleges were at the third level. There is no established technical basis for determining non-academic staff requirements;

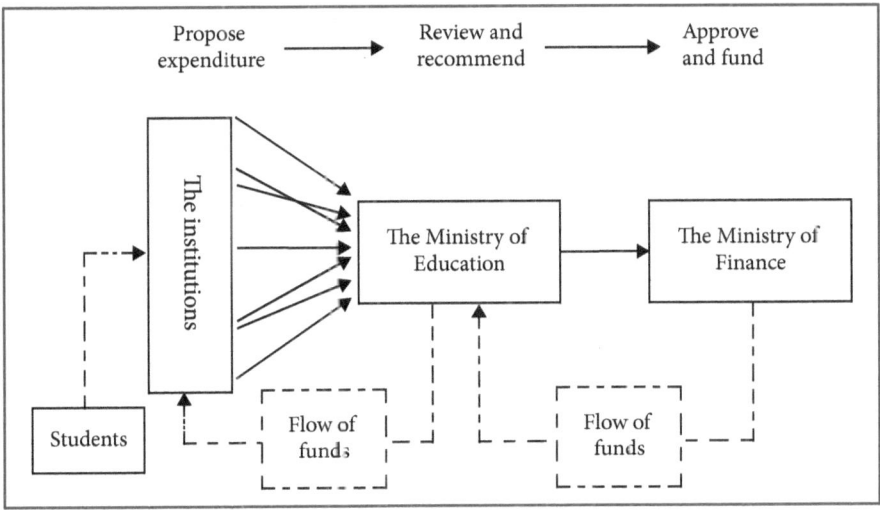

Figure 3.3. The resource allocation process for national higher education institutions in CARICOM

however, the academic staff numbers were theoretically determined by a student/staff ratio of 12.5:1. There was also no mechanism for reviewing the enrolment statistics with a view to adjusting the staffing numbers accordingly and this has resulted in the non-maintenance of the ratio. No funding is provided for research for the national tertiary level institutions.

With regard to fees charged to students, Jamaica's official policy is "cost sharing", where fees are charged to students to supplement the budget. In the case of UWI, it is the policy that the government pays 80 per cent of the economic cost of teaching and research, and the student fees cover the balance. Despite the official stance of the subsidy-fee ratio, there have been several cuts to UWI's budget over the past five years and, in the process, there has been no review of percentage subsidy to students. This suggests that the 80:20 ratio policy is not being maintained. It is important to note that UWI does not rely on this subsidy-fee structure to fund its operations; indeed, UWI funds up to 60 per cent of its operational budget independently from grants and other donor funds as well as full fee-paying students.

For the Jamaican national tertiary level institutions, the policy is that the student fees cover the non-staff expenses A review of the records of several national tertiary level institutions (UTech, Bethlehem Teachers' College and GC Foster Sports College) reveals that because of the lack of a process to review the enrolment with the aim of adjusting the staff numbers, institutions have resorted to charging fees to cover some of the staff costs. This also indicates non-adherence to the stated policy of state funds covering staff costs.

The Barbados Education Act, cap. 41, section 52, stipulated that public education in Barbados "shall be free", hence tuition fees are automatically covered by the government for students attending tertiary level institutions, which were recognized as public institutions. However, in August 2013, the Barbadian Government announced that, as of August 2014, tuition fees would be charged to UWI students. Tuition fees amount to 20 per cent of the costs. The government will continue to cover economic costs, while students will be responsible for tuition.

The document "Policy on Tertiary Education, Technical Vocational Education and Training and Lifelong Learning in Trinidad and Tobago" articulates the government's stance on financing post-secondary education by stating that "Government has expended more than US$2 billion on expanding capacity of the tertiary education and TVET system. This figure includes funding for public tertiary education and TVET institutions, skills training, and the provision of financial aid to students through the Government Assistance for Tuition Expenses (GATE) programme and the Higher Education Loan Programme (HELP)" (http://www.stte.gov.tt/Portals/0/Publications/Policy/Tertiary%20Education%20Policy.pdf).

Resource Allocation Methods in Caribbean Higher Education

Nkrumah-Young's (2010) resource allocation pendulum (see figure 3.4), which is used to assess the resource allocation models of the three Caribbean countries is explained, hereunder:

In the resource allocation pendulum debate, the transparency aspect addresses the manner in which the funds are determined for each HEI. It can either be negotiated or formulaic. In the negotiated system, each institution negotiates separately with the funding agency or Ministry of Education, as happens with Jamaica and Barbados. This process is politically driven and results in ad hoc arrangements. In the formulaic system, there are agreed formulae for determining the level of resources that are granted to the institutions. The formulae may be fixed or variable. The fixed system is where a government allocates a set percentage of the country's revenue to higher education (Ziderman and Albrecht 1995). Honduras is an example of fixed formula funding, where the constitution mandates that 6 per cent of the national budget goes to higher education. With variable funding, the amount may be determined by the level of input or output. An example of an input-based formula is that of Australia, which pays each institution by the number of enrolees multiplied by a rate per student. The output-based formula, on the other hand, is where the funding is granted according to the number of students who successfully complete the programmes of study. Denmark, with its taximeter, is an example of the output-based formula. On the transparency aspect of the pendulum, the Caribbean countries are

The Quality/Financing Conundrum in Caribbean Higher Education

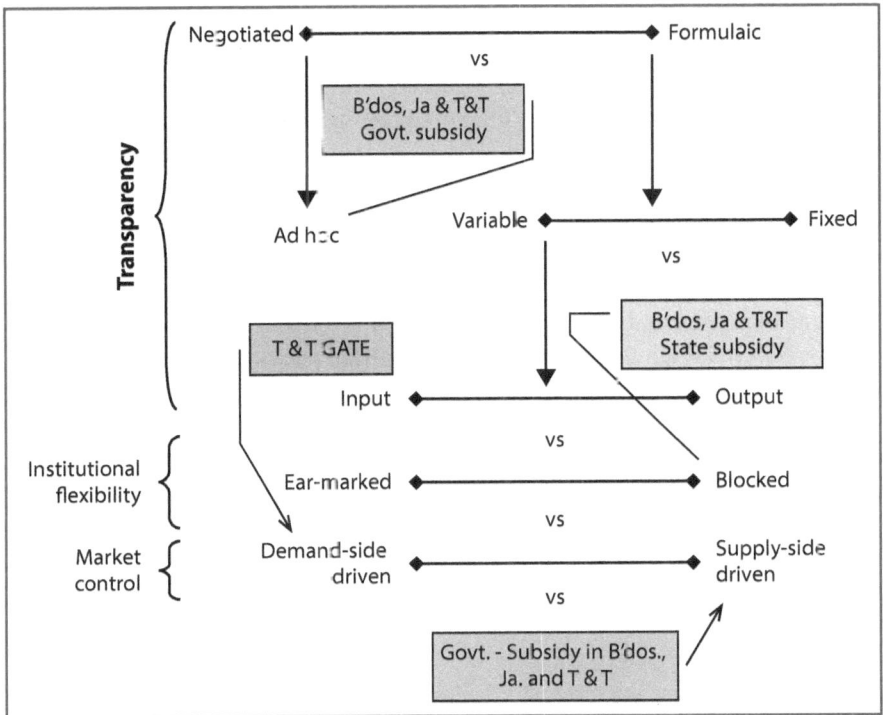

Figure 3.4. Resource allocation

considered to be at the ad hoc end. This is because of the separate negotiation process that each institution has to undergo for determining the resources received.

The institutional flexibility aspect addresses the authority of the HEI to use the funds as it determines. Within the earmarked system, funds are given for specific purposes and, if not used for such, then those funds would have to be returned to the authorities. With block grants, institutions are at liberty to use the funds as they choose. The Caribbean nations discussed practise block-grant funding as the institutions are at liberty to use the funds as they determine and are not required to return unspent portions.

The third element of the pendulum addresses market control. The market control debate concerns whether the funds are determined for the institution per se (supply side) or whether it follows the students (vouchers), according to their choice of institution (demand-side). In this area of assessment, Trinidad and Tobago is determined to be at the opposite end of the pendulum to Jamaica and Barbados.

Through the GATE programme, the state pays the fees for citizens and selected residents of Trinidad and Tobago attending tertiary institutions. In this system, social funding is provided for 100 per cent of tuition expenses for undergraduate students and up to 50 per cent, to a maximum of TT$10,000, for postgraduate students. All students who accept GATE funds are bonded for a period of national service, which is linked to the value of the funds received. National service is defined as employment within the Republic of Trinidad and Tobago in either the public or private sector (Report of the Standing Committee on the GATE Programme 2011). The GATE programme is determined to be a quasi-voucher system as the funding of the institutions through that method is demand-side driven. The student determines the institution to attend, whether public or private and, as long as the institution is recognized and the programme approved by the ACTT, students are eligible to apply for GATE funding. Since the tuition paid by the government is not a fixed sum applicable to all the institutions, then the system cannot be described as a total voucher system.

Barbados and Jamaica are determined to be at the supply side of the market control aspect of the pendulum where the funds are not determined by students' choice.

Assessing the Quality-Resource Allocation Link in Caribbean Higher Education

Using the Massy framework to assess the link between quality assurance and resource allocation to tertiary institutions leads to Massy's two questions: (1) Are the HEIs operating at or below the quality frontier? (2) Does the resource allocation model support the HEI's discretions on quality?

Grace Gordon, director of accreditation at the UCJ, points out that institutions and programmes are evaluated against established standards, and are at various stages of development along the quality journey (interview, August 22, 2013). The reports on evaluation visits that are sent to the institutions highlight the areas that are in need of improvement. In addition, the letters sent to the institutions indicating the UCJ's grant of registration and accreditation itemize the "terms and conditions" for the approved status; these terms and conditions are the areas to be addressed over the period of the registration and accreditation. Thereafter, the UCJ monitors the progress that is being made by the institution to address the areas of concern. The answer therefore to Massy's first question is that the institutions are at varied points in relation to the quality frontier. Some are closer to the frontier than others but some form of improvement is needed in all. The institutions, therefore, fall in column C of the Massy matrix.

With regard to Massy's second question, it was previously determined that there is no evidence that the negotiated ad hoc input resource allocation model in use in the Caribbean had any direct link to performance or the attainment of quality standards. Therefore the institutions are in the first row of the matrix.

The (C1) position on the matrix elicits three reactions about the Caribbean's quality/financing nexus. First, the institutions that are closest to the quality frontier may not need much incentive for high-quality performance but would lack the opportunity for reinforcing the commitment. Second, those that are far below the quality frontier may not be encouraged to perform better and improve their quality standards. Third, there is need for some form of encouragement to address the areas for improvement as indicated in the accreditation reports. The stratification of the salary element in the Jamaican education system, however, raises the question of the quality-trap as the institutions that are higher on the stratum would end up attracting the better staff thereby reducing the chances for improvement for those at the lower end.

Orr (2005) argues that funding provides a steering function for the state, while the quality assurance mechanisms are used for mapping or explaining the performance of the universities. As such, he deems that quality assurance mechanisms provide information to judge the effects of the funding. The analysis shows that the method of allocating funds to Caribbean tertiary institutions fall in the Orr's "discretionary budget" category. Jamaica's and Barbados's recognition process is voluntary and as such their quality assurance mechanisms fall in Orr's "marketing" category. The resource allocation systems in those countries do not correspond to their quality assurance mechanisms and this explains why the information produced by the quality assurance mechanism is not directly used to demonstrate the effectiveness of the funding.

The GATE system in Trinidad and Tobago, where, in order to receive such funding, the programme and institution must be recognized by the ACTT, results in the quality assurance mechanism being in the input category on Orr's trajectory. In this case, the resource allocation mechanism is supported by the quality assurance mechanism as the latter is used to justify the former. Orr explains that if the accreditation system is compulsory, as in the case of compulsory recognition either through registration or accreditation in Trinidad and Tobago, then the state uses it as a method of controlling input as input standards would be set for the institutions to achieve (email, May 23, 2005). If the accreditation system is voluntary, as in the case of Jamaica, then the institution will choose to take part so that it can better position itself in the market. The institution could also use it as a benchmark to improve internal performance but also as a marketing tool. Analysis of the Jamaica case further reveals that the incentive for institutions to be registered or accredited is driven by the demand of students, who need to be assured of value for money (fees) and the need to have a smooth transition to further education or employment in other countries. The positioning of the countries of the Caribbean is illustrated in figure 3.5.

With regard to Orr's second point on the use of the information from the mapping mechanism, it is observed that there is no deliberate use of the information

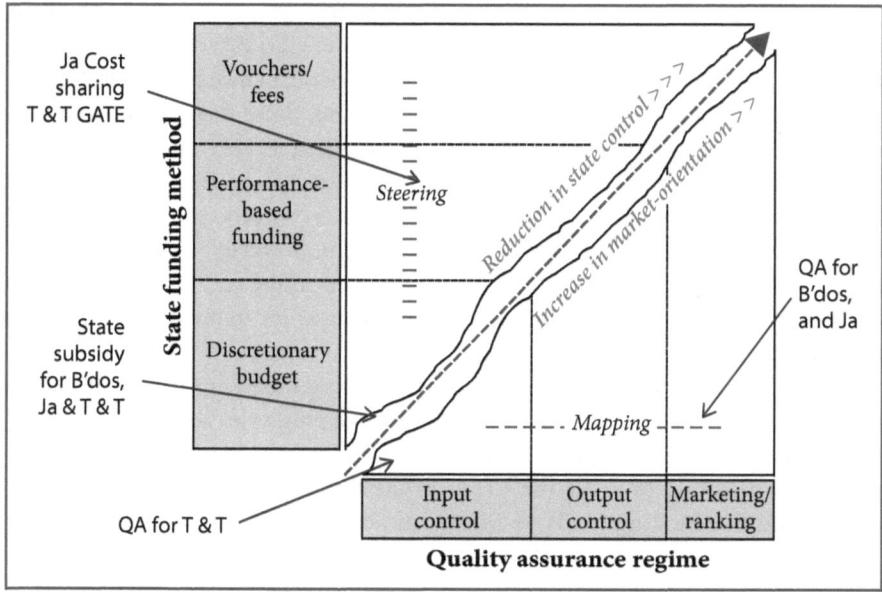

Figure 3.5. Positions on Orr

from the recognition (registration and/or accreditation) exercise in determining the level of funding by the respective ministries of finance. Ad hoc processes of determination, which often lead to duplication of efforts for verification, are requested through means other than reviewing registration or accreditation reports.

Conclusion

Massy believes that linking funding to performance provides incentives to maintain and improve quality and, in using his framework, it is revealed that for the Caribbean, the funding method has not served as an incentive for quality and could result in unsatisfactory returns on investment. Orr posits that funding could be a mechanism to implement or garner support for governments' policy; at the same time, the quality assurance mechanism could be used to provide information about the effects of the funding. Using his trajectory to analyse the relationship between funding of teaching and quality assurance, it is clear that for Trinidad and Tobago, the quality assurance mechanism, being compulsory in order to receive GATE, falls in the input control category thus correlating to the resource allocation mechanism of discretionary budgeting. In such a case, the quality assurance mechanism provides information for the resource allocation method. In the Jamaican and Barbadian cases, however, state funding is not linked to the quality assurance mechanism. Thus, those governments do not have systems in place to communicate the effect of their funding. The market element with fees being charged in Jamaica provides the link between the institutional quest for funds and the desire to be recognized by an external body.

In this chapter, analysis of the Caribbean situation was limited to funding and quality assurance of teaching. This limitation was necessary because funding for research was only provided to UWI and allocated in a way that does not distinguish it from that for teaching. Also there was no identifiable mechanism to assess the quality of research. The quest for efficient use of resources, however, suggests that there ought to be a more direct link between the method of allocating state funds to higher/tertiary institutions and the method of providing information on the impact of those funds on the quality of teaching as well as research output.

References

Burke, J.C., and M. Shahpar. 2000. "To Keep or Not to Keep Performance Funding". *Journal of Higher Education* 71 (4): 432–53.
Creswell, J. 2003. *Research Design: Qualitative, Quantitative and Mixed Methods Approaches*. 2nd ed. Thousand Oaks, CA: Sage.
The Future of Higher Education. 2003. British Government White Paper CM5735. Department of Education and Skills.
Harvey, L., and P. Knight. 1996. *Transforming Higher Education*. London: Society for Research into Higher Education, Open University Press.
Layzell, D.T. 1998. "Linking Performance to Funding Outcomes for Public Institutions of Higher Education: The US Experience". *European Journal of Education* 33 (1): 103–11.
Massy, W. 1996. "Value Responsibility Budgeting". In *Resource Allocation in Higher Education: The Economics of Education*, edited by W. Massy, 293–323. Ann Arbor: University of Michigan Press.
Mishra, S. 2006. *Quality Assurance in Higher Education: An Introduction*. Bangalore, India: National Assessment and Accreditation Council (NAAC).
Nkrumah-Young, K. 2005. "Exploring Financing Options for Higher Education in Jamaica". PhD diss., University of Bath.
———. 2010. "Reforming the Financing of Higher Education: Implications for Caribbean Administrators". *Journal of Education and Development in the Caribbean* 12 (1): 1.
Orr, D. 2005. "Can Performance-Based Funding and Quality Assurance Solve the State vs Market Conundrum?" *Higher Education Policy* 18: 31–50.
Report on the Standing Committee on the GATE Programme. 2011. *National Consultation: Securing and Expanding the GATE Programme*. Port of Spain: Ministry of Science, Technology and Tertiary Education.
Ziderman, A., and D. Albrecht. 1995. *Financing Universities in Developing Countries*. London: Falmer.

4 | Science, Technology and Innovation
Entrepreneurial Universities for Caribbean Development

PATRICK S. DALLAS AND G. JUNIOR VIRGO

The United Nations millennium development goals, an "international standard of reference for measuring and tracking improvements in the human condition in developing countries" (UN Task Force 2005, 2), offer a multidimensional development framework and a set of clear quantifiable targets to be achieved by 2015. Among the goals to be achieved are: a reduction in absolute poverty, malnutrition, hunger and universal primary enrolment and combating HIV and AIDS, halting and reversing the incidence of malaria and tuberculosis, access to reproductive health, and provision of safe drinking water and basic sanitation (CDCC 2006).

Meeting the millennium development goals requires a reorientation of development policies to focus on key sources of economic growth, including those associated with the use of new and established scientific and technological knowledge, along with related institutional adjustments. Furthermore, the fact that we live in a modern technological age also requires a population that is well educated and trained in science and technology (S&T).

Innovation (I) is another key factor in the development process, as it pays attention to the application of S&T to ever new problems while simultaneously improving solutions and products. It speaks to inculcating a mindset among citizens of constant improvement and entrepreneurship – the pursuit of a competitive edge – in the quest for development. Innovation's focus on constant improvement of process and solutions is akin to the notion of quality assurance and enhancement, which focuses on continuous improvement.

Today, we are experiencing a period of transformation and globalization at a pace never before experienced, accelerated by the application of science and new knowledge. As scientific and technological advances continue to transform our world, the Caribbean, too, is becoming increasingly aware of the importance of science, technology and innovation (STI) in the development process; and, in this regard, the region should also expect its universities, as leaders in science and technology education, research, and innovation, to play an important role in economic development by transforming into entrepreneurial universities in partnership with government and industry.

STI: How They Differ and How They Come Together

Betz provides the following definitions: "Science is the *discovery and explanation* of nature. Technology is *knowledge of the manipulation of* nature for human purposes. Nature is the *totality* of the essential qualities of the experienciable phenomena of the universe" (2003, 4).

Thus, Betz asserts, "Science *understands* nature, and technology *manipulates* nature" (ibid.). In providing a precise definition for innovation, he states, "Technological innovation is both the *invention* of a new technology and its *introduction* into the marketplace as a *new high-tech* product, process, or service" (22; emphasis added). And, in making the distinction between *invention* and *innovation*, Betz notes that "Invention is the creation of a functional way to do something, an idea for a new technology; innovation is the commercialization of new high-tech products, processes, or services" (ibid.).

Thus, we see here that the connection between science and economy is through technology, and technological innovation is simply a business process.

The Caribbean and STI: A Historical Perspective

While Caribbean Community countries are at different stages in their development, they fall largely in the group categorized as developing countries, and the region is seen as being in the global South. Like most global South countries, it is generally accepted that Caribbean Community countries do not spend enough on developing their STI capabilities. For example, Nurse (2007) argues that with regional research and development (R&D) expenditures estimated at an average of only 0.13 per cent of gross domestic product, Caribbean countries invest too little in R&D. This is low even for countries in the South, and a regional policy framework (CCST 2007) recommended increases to achieve an average target of at least 3 per cent of gross domestic product in order to attain "a median investment level between that of other Small Island Developing States, and that of developed countries" (5).

Despite the above, the Caribbean region has still managed to make a number of important contributions to the development of science and technology. A few illustrations from Jamaica – which, historically, has played a leading role in the region in the application of STI in national development – bear this out (Lowe, Magnus and Brown 2000). It is well known, for example, that the Jamaican capital, Kingston, had gas lighting (1877) and the town of Black River received electricity (1893) before New York City. Also, the town of Falmouth, Trelawny, had piped water in homes before urban New York City. Indeed, Falmouth's water supply system was one of the first ever piped water supply systems in the Western Hemisphere. The Martha Brae River was the water source for the Falmouth Water Works Company established in 1799 to supply the town and visiting ships (www.nwcjamaica.com). And a steam-driven sugar mill – invented by John Stewart, a millwright – was installed on a Jamaican sugar plantation in 1770, just seven years after steam power was first used in Britain. Jamaica was not only the first country in the Americas to build a railway (commissioned in 1845), but it also led the way in to using research results to improve sugar cane production. The telephone was already available in the offices and residences of some Jamaican officials by 1878, only two years after it was invented, and commercial telephone service became available in Jamaica in 1883.

During the 1880s, John Redman Bovel's research aided the cane breeding programme in Barbados by replacing the Bourbon strain of sugar cane with new varieties that gave much higher yields and unprecedented revenue for the troubled sugar industry (Galloway 1996).

More recently, Oliver Headley's passion for harnessing renewable energy technology has served to promote his native Barbados to the forefront of solar energy application within the Caribbean (CCST 2007). Indeed, the proliferation of solar crop dryers (Headley, Harvey and Osuji 1986) and solar stills (Headley and Springer 1971) in the small island state are but mere testaments to his untiring efforts.

Pioneering work by the Jamaican Thomas Lecky led to the development of the Jamaican Hope (declared a breed in 1952) and other cattle, which would thrive in tropical conditions, while Professor Manley West and Dr Albert Lockhart are credited with the development, in 1987, of a treatment for glaucoma from the marijuana plant called Canasol. Early work on Canasol started in the labs at the College of Arts Science and Technology (now the University of Technology, Jamaica [UTech]). Also, efforts led by Gerald Lalor resulted in installation of the SLOWPOKE research nuclear reactor on the Mona campus of the University of the West Indies (UWI) in 1984. In 1991, Jamaica was one of the few countries having a 100 per cent digital telephone network. Six years later, in 1997, Dr Paula Tennant, also of the Mona campus, became the first person in the Caribbean to produce a bio-engineered product, a transgenic strain of papaya. Noteworthy too is that following on a number of important developments in its bauxite/alumina industry over the years – in particular, solutions to the technical problems of precipitating the fine mud particles in the local bauxite industry, the high phosphorus content of Jamaican bauxite and the disposal of the red mud by dry stacking – Jamaica broke ground in February 2013 for the construction of a pilot plant to extract rare earth metals (lanthanides) from the red mud by-product of bauxite mining activities.

Significantly, too, Jamaica was among the earliest developing countries to promulgate legislation to guide the use of S&T for the exploitation of domestic natural resources (Jamaica S&T Policy 2005). Successive Jamaican governments have identified science and technology as priorities for development and have put in place various policy measures to support and encourage the development and utilization of science and technology. Consequently, more recent policy development has also sought to strengthen the role of innovation and the proper funding of S&T activities to stimulate diversification of agriculture, revival of manufacturing and expansion of services; as well as promote greater private sector participation in S&T development and application, while reducing poverty and unemployment (*National Development Plan: Vision 2030 Jamaica*).

STI Policies in the Caribbean

Concomitant with an increasing recognition of the need for STI policies to meet the challenges of the global environment and to enhance the existing human resources and institutional capacity for innovation and research and development, Caribbean policymakers are now showing more interest in exploring how STI policies can be used to upgrade the region's industrial and export capabilities (Nurse 2007). However, while it is the stated policy of all Caribbean governments to promote S&T, it is not always easy to judge how effectively this is being done (EUCARINET 2012). In the following paragraphs, we will look at some of the more recent attempts by the region to develop STI policies.

Aiming to integrate and harmonize national policies across the region, Caribbean Community heads of government agreed in 1988 to adopt a regional science and technology policy. The Caribbean Council for Science and Technology – a Caribbean umbrella organization for various national councils for science and technology – was then designated in 2000 as the agency to coordinate and implement the policy. A regional policy framework for action (*Science, Technology and Innovation for Sustainable Development – Caribbean Regional Policy Framework for Action*) was formulated and published in 2007 (CCST 2007). The rationale for the policy framework states:

The Caribbean Regional Policy Framework for Action provides a framework for guiding the formulation of decisions, and the consolidation of national and regional efforts, geared towards strengthening economic growth and diversification, by promoting, advancing, and sustainably utilizing STI. Such efforts have the capacity to increase economic productivity, to enhance competitiveness, to create wealth, to alleviate poverty and unemployment, and to provide practical solutions that may improve the quality of life for all residents of the Caribbean. The adoption of this regional policy framework for STI provides an opportunity for Caribbean governments to acknowledge the potential benefits that STI may provide, when regarded as national investments. (CCST 2007, 1)

The regional policy framework identifies the following nine supporting institutions and mechanisms and twelve priority policy areas:

Supporting institutions and mechanisms
1. infrastructure
2. policy and planning
3. development financing
4. innovation and entrepreneurship
5. standardization
6. human resource development
7. science and technology education
8. research and development
9. regulatory framework and intellectual property rights

Priority policy areas
1. agriculture and the food sector
2. biotechnology and biosafety
3. environmental management
4. coastal and marine resources management
5. waste management
6. integrated water resources management
7. alternate energy and energy management
8. disaster preparedness
9. health
10. sustainable tourism
11. development of small, medium and micro enterprises
12. information and communications technology

The overall expectation then is that STI will not just be given token recognition as important elements in economic and social transformation, but that their roles will now be generally integrated into the development planning process. This is reflected in the national development plans of major Caribbean Community countries such as Barbados (NDPB 2005), Jamaica (NDPJ 2009) and Trinidad and Tobago (NDPTT 2010). Notably then, Jamaican S&T policy, for example, now also reflects the importance of creating greater abilities and capacities to compete in global trade, protecting the island's delicately balanced ecology and biodiversity, as well as developing and maintaining scientific skills critical for implementing the policy (NDPJ 2009).

Universities as Partners in the STI Imperative

The importance of universities in setting the R&D agenda has been confirmed by several studies (Barge-Gila, Santa-María and Modrego 2009; Mansfield 1980; Mohnen and Hoareau 2003). Some of these studies also present evidence that there are high spillover effects from academic R&D (Adams 1990; Jaffe 1989). Acs, Audretsch and Feldman (1992) even found that the geographical proximity to universities increases innovation, in terms of both the degree of patenting and the number of new products introduced into the market. Silicon Valley in the southern region of the San Francisco Bay Area in northern California, United States, is perhaps the best-known example of this. Silicon Valley "grew up" around Stanford University, with its entrepreneurial leadership, especially in engineering and partnerships with private companies like Bell Telephone. The US government was also a player in the development of Silicon Valley, as the San Francisco Bay Area had long been a major site of US Navy research and technology.

Henry Etzkowitz (2008) summarizes the importance of university – industry – government partnerships thus: "The interaction among university, industry, and government is the key to innovation and growth in a knowledge-based economy" (1). Etzkowitz, in putting forward his *triple helix* representation, states that the university is the generative unit in knowledge-based organizations, just as government and industry were the primary institutions in industrial society. Today industry continues to be a key player as the locus of production; government as the source of contractual relations guarantees stable interactions and exchanges. Universities maintain a competitive advantage with the students they attract, since, unlike the R&D units in firms and government labs, each batch that matriculates is a source of new ideas. In a triple helix interaction, the university, firm and government take on each other's role while maintaining their own role and distinctiveness. The university takes on the role of industry by stimulating the development of new firms from research, "introducing 'the capitalization of knowledge' as an academic goal" (Etzkowitz 2008, 1). Firms develop increasingly higher levels of training and share knowledge through joint ventures – similar to the way universities function. Governments behave like venture capitalists without giving up their regulatory activities. Unlike other theories, which focus on government or industry as the drivers of innovation, the triple helix theory focuses on the university as the source of new technology, entrepreneurship and critical inquiry. Entrepreneurial universities play a key role in the triple helix by facilitating technology transfer, incubating new firms and taking the lead in regional renewal efforts.

Etzkowitz characterizes innovation as "the reconfiguration of elements into a more productive combination" (4). This takes on a broader meaning in increasingly knowledge-based societies. In recent years, the appropriate configuration of the relationship between firm-formation, high technology and economic growth has become a matter of public concern and much debate. Thus, some people are beginning to question whether the university is losing its traditional role and independence by becoming more closely involved with industry and government. The presumption is that the university is subordinate in that relationship. Or is it that universities are acquiring higher status through these reconfigured relationships and increasing their influence in society? In that regard, are universities enhancing their status in society and taking on a more central role by their contribution to innovation? Etzkowitz answers thus: "Knowledge-based economies are more tightly linked to sources of new knowledge; they are also subject to continuous transformation instead of being rooted in stable arrangements Fostering a continuous process of firm-formation based on advanced technologies, often university-originated, moves to the heart of innovation strategy"(4).

Caribbean Universities: Their Role in Setting the STI Agenda

The current information and communications technology age has widened the gap between the rich and the poor, and the developing and the developed, virtually forcing the developing nations, including the nations of the Caribbean, to question and critically examine the role of education in their development, their educational systems and the content of their curricula in advancing their nations. The information and communications technology revolution has undergirded the reform of education in virtually every single Caribbean country in the recent years.

Undoubtedly, universities and other tertiary level institutions have a key role to play in providing education that guarantees the kind of sustainable human development that the region requires. Caribbean universities have immense potential to promote technological development through the training of educators, thinkers, policymakers, technologists, technicians, craftspeople, innovators, technical and education specialists. They can also integrate into the productive sector and society at large in several ways: through conducting research that impacts directly on industry and business; involvement in capital formation projects, such as business incubators and technology parks; introducing entrepreneurial training and encouraging students to take research from university into business. Importantly, Caribbean universities can ensure that students study the relationship between science, technology, innovation and development; so that they are sensitive to societal needs and are motivated to respond.

In Jamaica, the three major universities – the University of the West Indies, the University of Technology (Jamaica) and Northern Caribbean University – have been important partners in the development of STI as a national imperative. All three universities have been carving out niches as they continue to make important strides in S&T, concomitant with increased appreciation of their own key role in steering national development through the harnessing and promotion of knowledge from scientific research and application.

Interestingly, though, while the universities are singularly suited to contribute to the task of promoting development through the application of STI, most Jamaican students – just like their contemporaries in other Caribbean islands – still pursue first and advanced degrees in the arts and humanities, and the social sciences. Some will argue that this is one of the reasons for the failure of Caribbean education to keep pace with the West. Enrolment data from the UWI, UTech and Northern Caribbean University suggest that the colonial influence on Caribbean education is still fairly strong; consequently, the mix of university education continues to be biased against knowledge-based disciplines. Student enrolment figures still show a strong preference for the professions, the teaching and public services, and other "white collar" jobs – jobs that gained status due to their long tradition in the Caribbean. Some figures bear examination.

At UTech – an institution started in 1958 with the explicit mandate to fuel national development objectives – enrolment in engineering and computer science in 2005 was only 28 per cent (UNCCA 2005). By 2010, with the inclusion of health sciences and other technology-based disciplines, this figure increased to 44 per cent. Intriguingly, for the 2011–12 academic year, enrolment in science and technology-based disciplines totalled 4,761 or 42 per cent of total enrolment; enrolment in computer science and engineering is at a low of 19 per cent. This is similar to trends in other developing countries and requires some serious interrogation.

Enrolment trends at the UWI Mona campus tell a similar but different story. For the 2005–6 academic year, 15,398 students were enrolled, with the majority – some 7,029 or 47 per cent – enrolled in the Faculty of Social Sciences. Humanities and Education followed with a total of 4,398 individuals or 29 per cent. These two faculties – Social Sciences and Humanities – together accounted for over three-quarters of the student enrolment (76 per cent). In contrast, Pure and Applied Sciences (now called Science and Technology) had only 2,080 (14 per cent) while Medical Sciences totalled 1,782 or 10 per cent. Thus, Pure and Applied Sciences and

Medical Sciences accounted for 24 per cent of the UWI Mona enrolment in 2005–6. For the same period, the Cave Hill campus in Barbados had total enrolment of 6,167 students, of which 17 per cent opted for either Pure and Applied Sciences or Medical Sciences. Notably, too, in 2005–6, UWI's Faculty of Agriculture at the St Augustine campus in Trinidad had an enrolment of 163 students, or about 1.1 per cent of the student body, and the Faculty of Engineering had an enrolment of 919 students, or about 6.2 per cent of the student body.

Of more recent date, in 2010–11 the profile at UWI Mona had changed somewhat, with a total enrolment of 11,466, of which 40 per cent were enrolled in Pure and Applied Sciences and Medical Sciences. For 2011–12, there was a further increase to 42 per cent, which suggests that UWI Mona, through the introduction of new technology-based undergraduate and postgraduate programmes within the last few years – for example, the BSc in Electronics Engineering, BSc in Tropical Horticulture and MSc in Entrepreneurship in Agriculture – has started to take S&T more seriously. Other technology-focused programmes, such as information technology, agro-processing, financial mathematics, enterprise risk management, civil engineering, energy and power engineering, computer engineering, and biomedical engineering, have also been introduced over the past two years. Furthermore, the recent name change of the Faculty of Pure and Applied Sciences to the Faculty of Science and Technology underlines this new technology-focused direction.

The trends in enrolment profiles at the universities in Jamaica, as noted above, can be interpreted as an indication that the literary and traditional bias in education inherited from colonial times has begun to change. Even so, the change is slow, and the enrolment numbers for students who show a preference for the professions with a desire to harness scientific knowledge for productive use are still low. The educational systems in Jamaica are therefore still not producing an adequate pool of scientific or knowledge workers needed to make the country globally competitive.

Interestingly, the conundrum highlighted above is not limited to the Caribbean region. Indeed, while the scientific, technological and engineering community – together with their associated institutions, such as universities, technical institutes and professional bodies – continues to be highlighted as being among the most critical resources for economic transformation, a disturbing global trend is the decline in enrolment in engineering courses, especially in developing countries. As if this were not bad enough for the developing countries, more developed countries are constantly recruiting from developing countries in order to meet their shortage of scientists and engineers. As the UN Task Force Report sums up the dilemma: "Ironically, developing countries are putting their scarce resources into education and training that benefits the developed world" (Task Force 2005, 91). Thomas-Hope (2003) wrote that the education levels of migrants from the Caribbean usually average higher than those of both their homeland and the country accepting them.

Gallina (2010) also reported this practice, which he recognized as a regional trend, and termed it "irregular migration". Jamaica, it should be noted, exports about 80 per cent of its tertiary-trained graduates (Gregory 2013).

The STI Imperative: New Roles for Caribbean Universities

Undoubtedly, Caribbean universities must continue to evolve and mirror more closely those of the more developed countries, taking on ever greater importance and prominence in the development process. In this evolution, the universities have to be more than mere degree- or certificate-granting institutions, providing the documentation for their graduates to apply for jobs. Further, in addition to their very important, but more obvious role in training and inculcating students in technical skills and critical thinking, the universities need to continue (and even accelerate) the process of introducing students to the relationships between science, technology, innovation and development, so that they are sensitive to societal needs.

In another development, universities are increasingly being viewed as valuable resources for business and industry. Caribbean universities, too, have to embrace the need to get involved in undertaking entrepreneurial activities with the objective of improving regional and national economic and social performance. As the universities seek to integrate in the productive sector and contribute to economic development, they must continue developing their expertise and facilities to allow them to undertake R&D assignments for industry, encourage innovation and entrepreneurial endeavours, as well as provide technical advice to guide policy development and support government initiatives.

The Entrepreneurial University

In order to keep pace with their more developed counterparts, Caribbean universities must also embrace Etzkowitz's idea of having "the capitalization of knowledge" at the heart of their mission on the evolutionary path towards transformation into the entrepreneurial university, where they become established as economic actors in their own right. Thus, they need to start giving more serious attention to putting in place the policies and strategic approach to building the four pillars on which Etzkowitz's entrepreneurial university rests (2008, 27):

1. Academic leadership able to formulate and implement a strategic vision (with considerable degree of independence from state and industry).
2. Legal control over academic resources, including physical property such as buildings and intellectual property emanating from research.

3. Organizational capacity to transfer technology through patenting, licensing and incubation.
4. An entrepreneurial ethos among administrators, faculty and students.

The entrepreneurial Caribbean university must now move to mine research findings for their technological potential and translate them into use. The university is a natural incubator since it provides a support structure (time, social and physical space) for students and faculty to initiate new ventures (political, intellectual or commercial). Such new ventures are exportable across highly permeable borders. The university is a seedbed for new scientific fields and new industrial sectors, each cross-fertilizing the other. Biotechnology is a recent example of this phenomenon as pharmacology was in the seventeenth century.

Beyond these natural entrepreneurial features, a university that is entrepreneurial takes the lead in putting knowledge to use and broadening the input in the creation of academic knowledge. Etzkowitz notes the following: "The university must identify the areas of research and teaching that it will focus on to create 'steeples of excellence' in order to attract significant support and external funds. An entrepreneurial university also has the capacity to take in and address problems and needs from the larger society, making them the basis of new research projects and intellectual paradigms, creating a virtuous circle with internal intellectual development" (28).

University of the West Indies and University of Technology (Jamaica)

Encouragingly, we are beginning to see signs of Caribbean universities taking on more of the characteristics of the entrepreneurial university. This is reflected, for example, in UWI establishing the Office of Sponsored Research to protect intellectual property and, among other things, manage the transfer of technology through patenting and licensing. In addition, UTech has created a technology park and a business incubator, and all three major universities in Jamaica have created spin-off firms in the form of research institutes, to build capacity in R&D and problem-solving, and present a vehicle that enhances the working relationship with industry. These include the Biotechnology Centre and the Caribbean Genetics (CARIGEN, the DNA testing centre), both at UWI Mona. The UWI Cardiac Surgery Simulator developed by Drs Paul Ramphal (Medical Sciences) and Daniel Coore (Computing) highlights the commercial potential of inventions. It has garnered significant interest among professionals in the field and cardiac equipment industry personnel; UWI does not have the infrastructure to produce the simulator but will partner with investors to do so. Over the past thirteen years, UWI has held research days to promote the research and innovation of its campuses.

Northern Caribbean University

Northern Caribbean University, a privately owned institution, in contrast to its sister institution Southern Caribbean University in Trinidad and Tobago, in a short time, has developed an Office of Research and Publications as the primary agent of its current thrust to foster research. In tandem with this Northern Caribbean University developed a policy on intellectual property. Among the various initiatives, Northern Caribbean University has garnered global respect in the world of information and communications technology with regional and global wins at Microsoft's Imagine Cup and the Digital Jam 2.0 project in 2010. They won the Blackberry Developers Competition in 2011 for their Mathberry application. In March 2012, the university, in partnership with Scotiabank Jamaica, opened up a new agricultural research centre to provide information on increasing crop yield, as well as laboratory and consultation services for farmers.

At the opening of the centre, the then CEO of Scotiabank, Bruce Bowen, stated:

> This lab further positions NCU as the centre of innovation and research and particularly for research that is relevant to the farming community in the areas immediately around the university. Mid-island Jamaica is the largest producer of agricultural produce in the country and it is therefore fitting that the research centre be based here. We look forward to the contribution and work that will be coming out of the facility in advancing the development of a modern, efficient and internationally competitive agricultural sector. (*Jamaica Observer*, March 25, 2012)

The university continues its research in areas such as the cancer fighting properties of garlic and sorrel.

University of Trinidad and Tobago

The University of Trinidad and Tobago identifies as its mission: "To be an *entrepreneurial* university designed to discover and develop entrepreneurs, commercialize research and development, and spawn companies for wealth generation and sustainable job creation towards the equitable enhancement of the quality of life of all individuals, families and communities of the Republic of Trinidad and Tobago and the Caribbean" (UTT website).

In light of its mission, University of Trinidad and Tobago (UTT), the national university of Trinidad and Tobago, continually refers to itself as an entrepreneurial university and attempts to be active in R&D, especially in engineering and information and communications technologies. UTT acting president and provost Fazal Ali in 2012, at the signing of a memorandum of understanding (MOU) with CrimsonLogic PTE Ltd, a Singapore-based provider of eGovernment solutions said, "Human capabilities in ICT must be developed as a platform for economic development using talent, technology and tolerance as a basis to move ICT forward" (UTT website). UTT's approach to the relationship between entrepreneurship and research was on display in 2010, when it hosted a three-day entrepreneurial conference, "The Role of Universities in Entrepreneurship for Socio-Economic Development: UTT's Response". The featured lecturer on the first day was Henry Etzokowitz, the originator of the "Entrepreneurial University" and the "Triple Helix" concepts, discussed previously. Among the latest examples of the entrepreneurial drive is the May 2013 signing of an MOU with Toon Boom, the worldwide leader in digital content and animation creation software, to set up incubators in UTT's Animation Studies programme. Students are encouraged to start their own business, participate in a global, multibillion dollar industry, and earn a diploma at the same time.

Conclusion

STI underpin every one of the millennium development goals, and the scientific, technological, and engineering community and the associated institutions are among the most critical resources for economic transformation. Caribbean universities, as members of this community, have an important part to play in exploiting the potential of STI in national and regional development. This includes building local capacity that can help to solve several science and engineering-related problems that the region faces. In this regard, too, the universities are critical to Caribbean Community's efforts at achieving the millennium development goals and the aims of the various national development and strategic plans. Universities need to embrace their entrepreneurial potential to ever deepen their contribution to achieving the millennium development goals.

However, for the universities to have the best chance of positively influencing the development process, changes are required at several levels of administration. High school curricula need to be modified to prepare students for what they will encounter in university. Teaching methods must be changed to reflect the spirit of inquiry by encouraging independent projects, exposure to experts and hands-on experiences through internships and field trips. As Crowder et al. put it: "Scientific knowledge is changing very quickly as modern communication technologies facilitate the global sharing of information among scientists and educators. Since 'new' knowledge becomes 'old' knowledge so quickly, it is essential that graduates develop the skills and attitudes that will allow them to continue to learn and develop their competencies throughout their professional lives" (1999, 7).

It is also essential that Caribbean universities continue to embrace their evolving new roles, so that they can be more flexible and responsive to the demands of driving national and regional development. Universities and industry will have to be encouraged to work more actively together. Government also has a role to play.

References

Acs, Z., D. Audretsch and M. Feldman. 1992. "Real Effects of Academic Research: Comment". *American Economic Review* 82 (1): 363–67.

Adams, J.D. 1990. "Fundamental Stocks of Knowledge and Productivity Growth". *Journal of Political Economy* 98 (41): 673–702.

Barge-Gila, A., L. Santamaría and A. Modrego. 2009. "Complementarities between Universities and Technology Institutes: New Empirical Lessons and Perspectives". CÀTEDRA INNOVA Working Papers.

Betz, F. 2003. *Managing Technological Innovation – Competitive Advantage from Change.* 2nd ed. Hoboken, NJ: John Wiley & Sons.

CCST (Caribbean Council for Science and Technology). 2007. "Science, Technology and Innovation for Sustainable Development: Caribbean Regional Policy Framework for Action". *Caribbean Council on Science and Technology Report.* Newtown, Trinidad and Tobago. http://www.niherst.gov.tt/s-and-t/projects/ccst-sti-framework.pdf.

CDCC (Caribbean Development & Cooperation Commission). 2006. "Report on Science and Technology Infrastructure and Policy in Selected Member and Associated Member Countries of the CDCC", *Caribbean Development and Cooperation Commission Report.* UN Economic Commission for Latin America and the Caribbean (ECLAC), Subregional Headquarters, Port of Spain, Trinidad and Tobago. http://www.eclac.org/publicaciones/xml/5/22925/L.045.pdf.

Crowder, V.L., W.I. Lindley, T.H. Bruening and N. Doron. 1999. "Agricultural Education for Sustainable Rural Development: Challenges for Developing Countries in the 21st Century". FAO Research, Extension and Training Division, SD Dimensions, Rome.

Etzkowitz, H. 2008. *The Triple Helix: University-Industry-Government Innovation in Action.* New York: Routledge.

EUCARINET. 2012. "Regional Report on Priorities: Mapping of the Existing Caribbean Priorities in Science, Technology and Innovation". Report prepared by MENON Network, edited by APRE, Italy.

Gallina, A. 2010. *Human Mobility Report 2011: Migration and Human Development in ACP Countries.* Brussels: ACP Secretariat.

Galloway, J.H. 1996. "Botany in the Service of Empire: The Barbados Cane-Breeding Program and the Revival of the Caribbean Sugar Industry, 1880s–1930s". *Annals of the Association of American Geographers* 86 (4): 682–706.

Gregory, Robert. 2013. "Education for Home or Export: Policy Choice or False Debate?" *Gleaner,* February 3.

Headley, Oliver, and Basil Springer. 1971. "Distilled Water from Solar Stills". *Journal of Chemical Education* 48 (1): 49.

Headley, O., W. O'N. Harvey and P.O. Osuji. 1986. "Simple Solar Crop Dryers for Rural Areas". In *Proceedings of ISES Solar World Congress, 22–29 June,* Montreal, Canada, edited by E. Bilgen and K.G.T. Hollands, vol. 2, 1082–86. New York: Pergamon.

Jaffe, A.B. 1989. "Real Effects of Academic Research". *American Economic Review* 79 (5): 957–70.

Jamaica S&T Policy. 2005. "Science and Technology for Socio-Economic Development: A Policy for Jamaica". National Council on Science and Technology, OPM, Kingston.

Lowe, H., K. Magnus and Y. Brown. 2000. *Discovering the Future: The Emergence, Development and Future of Science and Technology in Jamaica*. Kingston: Pelican.

Mansfield, E. 1980. "Basic Research and Productivity Increase in Manufacturing". *American Economic Review* 70: 863–73.

Mohnen, P., and C. Hoareau. 2003. "What Types of Enterprise Forge Close Links with Universities and Government Labs? Evidence from CIS-2". *Managerial and Decision Economics* 24: 133–45.

NDPB. 2005. *The National Strategic Plan of Barbados 2005–2025*. http://www.sice.oas.org/ctyindex/BRB/Plan2005-2025.pdf.

NDPJ. 2009. *National Development Plan: Vision 2030 Jamaica*. http://www.vision2030.gov.jm/Portals/0/NDP/Vision%202030%20Jamaica%20NDP%20Full%20No%20Cover%20(web).pdf.

NDPTT. 2010. *Trinidad and Tobago Vision 2020 National Strategic Plan: 2008/2009 Progress Report*. http://www.finance.gov.tt/content/pubFAFC0C.pdf.

Nurse, K. 2007. "Science, Technology and Innovation in the Caribbean". Paper presented at Technology Policy and Development in Latin America, UNECLAC, Santiago, Chile, December.

Thomas-Hope, E. 2003. "Irregular Migration and Asylum Seekers in the Caribbean". Discussion Paper no. 2003/48. United Nations University World Institute for Development Economics Research, Helsinki.

UNCCA. 2005. "United Nations Theme Group Report on Education in Jamaica". United Nations Common Country Assessment Report, Kingston.

UN Task Force. 2005. "Innovation: Applying Knowledge in Development". In *UN Millennium Project, Task Force on Science, Technology and Innovation, 2005*, lead authors Celestous Juma and Lee Yee-Cheong. London: Earthscan.

2 | External Explorations

5 | A Critical Look at the Legislative Framework for External Quality Assurance Agencies in CARICOM

Ruby S. Alleyne

Globally, external quality assurance of higher education has three major aims. The first is to provide assurances of the quality of teaching, research and other activities of higher education institutions, for example, through accreditation. Secondly, the aim is to contribute to the improvement of higher education and, third, to provide reliable and verifiable data on the performance of higher education institutions. International developments such as the Washington Accord in 1989 and the Bologna Process in 1999 have served to highlight and strengthen the role that external quality assurance agencies (EQAAs) play in enabling greater transparency and comparability of standards across national boundaries; and also in facilitating increased mobility of skilled workers and the transnational recognition of qualifications. The Washington Accord is an international agreement among bodies responsible for accrediting engineering degree programmes in the signatory countries. It recognizes the equivalency of programmes accredited by the bodies and graduates of the programmes are recognized by other signatory bodies as having met the academic requirements for entry to the practice of engineering. The Bologna Process created the European Higher Education Area in 2010 with the main objective of ensuring that there are more comparable, compatible and coherent systems of education across Europe. Similarly, in the Caribbean, the establishment of the Caribbean Single Market and Economy (CSME) has served, among other things, as a catalyst for the emergence of a growing network of national quality assurance bodies across the region. It is envisaged that these bodies would play a pivotal role in facilitating increased intra-regional mobility of skills.

This chapter presents an overview of national frameworks for external quality assurance and accreditation in Caribbean Community (CARICOM). It examines areas of commonality and divergence in legislation, functions, governance and autonomy in decision making. Issues related to accountability and transparency and the imperatives for ensuring comparability of quality and standards in the regional context are also tackled.

Methodology

It is possible to compare the legislative frameworks for the establishment of national accreditation bodies in CARICOM; and in so doing, some areas of convergence and divergence emerge. Copies of the acts and other official documents were reviewed and the content was analysed. An online survey was also administered to the chief executive officer or chairperson of each national accreditation agency. The heads of the nine agencies that are currently in operation contributed to the survey (see table 5.2). Data from the survey served to support conclusions drawn from the documentary analysis and provided additional information on current operations. Survey findings were also circulated to the participating heads for verification. Using the findings from the research, the chapter examines the legal frameworks in eleven member states in which national legislation has been enacted; the functions ascribed to the national accreditation bodies in the acts; and the governance structure with respect to autonomy of decision making. It concludes by highlighting issues related to transparency and comparability of quality and standards across the region.

CARICOM and Regional Mobility

The CARICOM came into effect on 1 August 1973 by the signing of the Treaty of Chaguaramas. The fifteen member countries (Antigua and Barbuda; Bahamas; Barbados; Belize; Dominica; Grenada; Guyana; Haiti; Jamaica; Montserrat; St Lucia; St Kitts and Nevis; St Vincent and the Grenadines; Suriname; Trinidad and Tobago) and five associate members states (Anguilla; Bermuda; British Virgin Islands; Cayman Islands; Turks and Caicos Islands) are committed to regional integration, which facilitates improved standards of living, sustained economic, social and cultural development, and enhanced international competitiveness.

The Treaty of Chaguaramas was revised in 1989 and CARICOM embarked on the establishment of the CSME as "a single economic space within which business and labour operate in order to stimulate greater productive efficiency, higher levels of domestic and foreign investment, increased employment, and growth of intra-regional trade and of extra-regional exports" (Girvan 2007, 8). Article 46(1) of the revised treaty identifies "movement of Skilled Community nationals" as one of the key elements of the CSME and states that signatories have agreed:

To accord to the following categories of CARICOM nationals the right to seek employment in their jurisdictions: university graduates, media workers, sportspersons, artists and musicians, recognized as such by the competent authorities of the receiving member states.

To facilitate the achievement of this objective, emphasis was placed on removing obstacles to intra-regional mobility of skills. A deadline of January 2003 was originally set for the enactment of regulations to facilitate the movement of skilled graduates, but there were several impediments and the pace of implementation of the CSME lagged way behind early expectations (Wickham et al. 2004).

One of the major hurdles was the absence of agencies deemed as "competent authorities" to facilitate increased mobility of skilled persons by making legitimate judgements about the transnational recognition of qualifications awarded by institutions operating within the region as well as overseas. While formal systems for accreditation of higher education institutions had existed in countries like the United States for over a century, there was no regional mechanism for quality assurance and accreditation within CARICOM at the time that the CSME was established. The pace of development in the region in this regard had been slow. The University Council of Jamaica (UCJ), which was established in 1987, had been entrusted by the government of Jamaica with a broad mandate that included the quality assurance of tertiary level programmes. However, there was a need for a review of the legislation to provide, among other things, for the accreditation of institutions. In Trinidad and Tobago, a decision had been taken in 1971 to establish a national commission on accrediting (Cabinet Minute 1297 of 9 December 1971), but thirty years later this has not been achieved. In its place the Committee on the Recognition of Degrees, operating within the National Institute of Higher Education, Research, Science and Technology, carried out limited quality assurance activities.

Regional Accrediting Body

In the higher education environment at the regional level, the rapid expansion of higher education provision to meet increasing demand by out-of-school youth as well as adult learners seeking to advance their careers or engage in lifelong learning, presented several challenges. In spite of the expansion of the University of the West Indies, with three physical campuses and the Open Campus (offering online and distance education), student demand continued to exceed the available opportunities and this precipitated an increase in the number of public and private non-degree and degree-granting institutions operating in the region. In this context, there was a need for effective systems for setting standards and monitoring and evaluating the provision of post-secondary education at the national level. At the regional level, there was a desire to develop a mechanism to assess the equivalence of qualifications earned from institutions across member states. There was also an urgent need to implement measures to regulate cross-border initiatives such as the establishment of offshore universities and the increasing number of alliances between local institutions and foreign partners, several leading to the award of degrees by the overseas partner institutions.

In 2002, CARICOM (through the Council for Human and Social Development) took a decision to give priority to the establishment of autonomous national accreditation bodies by 2003 in furtherance of the objectives of the CSME. Draft legislation proposed by the CARICOM Secretariat provided a policy framework within which national bodies would have responsibility for "conducting and advising on the accreditation and recognition of educational and training institutions, providers, programmes and awards, whether foreign or national and for the promotion of the quality and standard of education and training" (National Accreditation Council [Agency] Bill No. 2002). The draft bill served as a template or guide which would set broad expectations for EQAAs in general while giving each member state the freedom to identify specific functions of the agency and define governance and administrative arrangements for its efficient and effective operations.

Legislative Framework

Eleven member states enacted legislation during the period 2004–11 to establish national accreditation bodies. Prior to this period, two countries, Jamaica and St Christopher (Kitts) and Nevis, had enacted legislation in 1987 and 1999, respectively. The fact that the legislation in these two countries predated the draft bill circulated by CARICOM in 2002, resulted in greater divergence between the legislative frameworks as established in the acts passed before 2004 and those passed thereafter.

In the case of Jamaica, while the survey data shows that the UCJ has an expanded mandate in accordance with government policy, the provisions of the act do not specifically refer to accreditation and other quality assurance functions. The St Kitts and Nevis legislation makes reference only to institutional accreditation. While the act was amended in 2001, this amendment served only to replace "full, open-ended accreditation" with "full accreditation reviewable after every five years" and to repeal the provision that "Any institution which is not accredited anywhere and does not wish to be accredited by the Government shall operate in accordance with existing laws relating to educational institutions" (section 7, subsection 3). This subsection was replaced as follows: "No institution shall operate in Saint Christopher and Nevis unless it is accredited by the Accreditation Board in accordance with the provisions of this Act." In the acts passed since 2004, only the Accreditation Council of Trinidad and Tobago Act (Act No. 16 of 2004) has since been amended, first in 2007 and then in 2008 (Act No. 16 of 2007 and Act No. 10 of 2008). Amendments to the legislation in Trinidad and Tobago sought to extend the transitional period, which allowed institutions to continue to operate while seeking to be registered by that council, and also introduced provisional registration for institutions that meet basic requirements but are not fully compliant with all of the standards for registration (Chapter 39:06). These measures were taken to ensure a smooth transition from a previously unregulated system to a regulated one.

While legislation has been enacted in Belize, the national body has not yet been established. In St Lucia, draft legislation is currently being reviewed. Table 5.1 provides details of the legislation passed in the Caribbean region and any subsequent amendments.

Most of the national accreditation bodies in CARICOM are still in the embryonic stage. Notably in Belize, although legislation was passed in 2004 and assented to on 7 January 2005, the board and the executive director were never appointed. Consequently the National Accreditation Council of Belize has not yet been established. As presented in table 5.1, the average time that elapsed in most countries between the enactment of legislation and the appointment of a board was two years. This is with the exception of Jamaica, Suriname, and Trinidad and Tobago where boards were appointed in the same year as the enactment of the legislation. The relatively slow pace of development with respect to putting a governance structure and administrative systems in place to make these bodies operational, has in large part stymied their development as they were not adequately or appropriately equipped to fulfil the wide mandate they were given with respect to quality assurance and accreditation.

Functions of National Accreditation Bodies

The EQAAs in CARICOM are charged with the responsibility for carrying out multiple functions, which, in developed countries with more resources and more expansive higher education sectors, would be performed by different agencies – both public and private. As a consequence of the limited resources available to the regional territories, it is essential that they concentrate their efforts on key strategic areas. Consistent with an analysis of the functions and powers of the agencies as stated in the legislation, the survey revealed that most of the EQAAs established in CARICOM since 2004 have been focusing their attention primarily on developing systems for the registration of institutions (which ensures that they meet minimum standards); institutional accreditation; assessment/recognition of qualifications

Table 5.1. Legislative framework establishing national accreditation bodies in CARICOM countries

Country	Legislation	Purpose of the act (as stated in the legislation)	Status of the Agency
Antigua and Barbuda	Act No. 4 of 2006	An act to provide for the National Accreditation Board, to vest, in that Board the power to accredit post-secondary institutions and programmes of study in Antigua and Barbuda and elsewhere, to provide for the process and mechanism of accrediting institutions and programmes of study and to provide for other related matters	Established. Board appointed in 2008
Barbados	Act No. 11 of 2004	An act to provide for the establishment of a body to be known as the Barbados Accreditation Council with responsibility for: (a) the registration of institutions offering post-secondary and tertiary education and programmes of study; (b) the accreditation of qualifications offered in Barbados; (c) the examination and verification of Certificates of Recognition of Caribbean Community Skills granted to Community nations by their own country for the purpose of giving effect to Article 46 of the Revised Treaty of Chaguaramas; and (d) the granting of Certificates of Recognition of Caribbean Community Skills in accordance with the provisions of the Caribbean Community (Movement of Skilled Nationals) Act, 2004; and for related purposes	Established. Board appointed in 2006
Belize	Act No. 20 of 2004	An act to provide for the establishment and incorporation of a body to be known as the National Accreditation Council of Belize, to vest that body with the power to grant recognition to awards obtained in Belize and elsewhere, including education delivered by electronic and other media; to determine the equivalency of all awards and certificates within the context of the regional qualifications framework; to take such measures as are deemed appropriate in order to ensure that all programmes and courses delivered in Belize meet the academic and professional standards required, and that educational quality is continuously enhanced; to provide service of public information regarding publicly and privately offered post-secondary education; to take measures and enter into agreements to facilitate regional mobility of skilled persons; and to provide for matters connected therewith or incidental thereto	Not yet established

Dominica	Act No. 13 of 2006	An act to provide for the National Accreditation Board and to provide for the process and mechanism of accrediting institutions and programmes of higher education in Dominica and other related matters	Established. Board appointed in 2008
Grenada	Act No. 15 of 2011	The accreditation bill seeks to provide for a National Accreditation Board and, for the process and mechanism of accrediting institutions and programmes of higher education in Grenada	Established. Board appointed in 2014
Guyana	Act No. 12 of 2004	An act to provide for the establishment of the National Accreditation Council to vest in that body the power to grant recognition to "awards obtained in Guyana and elsewhere, to determine the equivalence of all awards for the purpose of establishing acceptable standards within the Caribbean Community and for purposes connected therewith or related thereto"	Established. Board appointed in 2006
Jamaica	Act No. 23 of 1987 Act No. 16 of 1991	The function of the Council shall be to promote the advancement in Jamaica of education, learning and knowledge in the fields of science, technology and the arts by means of the grant of academic awards and distinctions and for that purpose: (a) to determine the conditions governing the grant of such awards and distinctions; and to approve courses of study to be pursued by candidates to qualify for such grants, including, where appropriate, arrangements for training and experience in industry or commerce associated with such courses	Established. Council appointed in 1987
St Christopher and Nevis	Act No. 21 of 1999 Act No. 9 of 2001	An act to provide for the process and mechanism of accrediting institutions in St Christopher and Nevis, and to provide for related matters	Established. Board appointed in 2002

(*Table 5.1 continues*)

Table 5.1. Legislative framework establishing national accreditation bodies in CARICOM countries (*continued*)

Country	Legislation	Purpose of the act (as stated in the legislation)	Status of the Agency
St Vincent and the Grenadines	Act No. 35 of 2006	An act to provide for the National Accreditation Board, to vest in that Board the power to accredit institutions and programmes of study in St Vincent and the Grenadines and elsewhere, to provide for the process and mechanism of accrediting institutions and programmes of study and to provide for other related matters	Established. Board appointed in 2008
Suriname	Act No. 74 of 2007	Having taken into consideration that, with reference to advancing the quality of higher education and guaranteeing the regional and international comparability of courses in higher education, it is advisable to establish a national organization for accreditation and a central register of courses	Established. Board appointed in 2011
Trinidad and Tobago	Act No. 16 of 2004 Act No. 16 of 2007 Act No. 10 of 2008	An act to provide for the establishment of an Accreditation Council of Trinidad and Tobago and for related matters	Established. Board appointed in 2004

earned outside of national boundaries; and facilitation of free movement by qualified nationals in the CSME, by the issuance of a "skills certificate" giving approval for their qualifications to be accepted by employers in another CARICOM country. Few of the agencies have been engaged in activities related to the broader mandate, which emerged in the draft CARICOM bill. These include:

- programme accreditation
- conferment of institutional title or regulation of degree-granting powers (for new institutions seeking to establish themselves as universities or use other protected titles)
- recognition of foreign or transnational institutions and their awards
- assessment of the equivalence of qualifications
- evaluation and validation of new programmes
- development of unified credit-based systems for their higher education sectors

The provision of multiple services presents significant resource challenges for the EQAAs at the early stages of their development. Staff size varies and all of the EQAAs rely primarily on the state for funding. The draft bill provided for the EQAA to "fix and collect fees in connection with the exercise of its function" (section 4 [3] [n]). This was adopted by all territories in their legislation, which also gives them the power to borrow funds. The revenue-generating capacity of the recently established bodies, however, is directly linked to the extent to which they can offer the full range of services provided for in the act. Currently, institutions seeking registration or accreditation pay fees that cover the expenses incurred by the EQAA in carrying out the assessment. It is also envisaged that funds can be raised through hosting education and training events such as workshops and seminars for faculty and staff of tertiary level institutions. In this context, EQAAs will continue to be heavily dependent on state funding in the foreseeable future, while strengthening institutional capacity to offer services that are self-financing and revenue generating.

Based on the survey findings, the Accreditation Council of Trinidad and Tobago has a full-time staff of fifty-six; the UCJ has twenty-eight; the Barbados Accreditation Council has seventeen; at the other EQAAs, full-time employees generally number between one and three. Although most EQAAs rely on a pool of qualified and trained persons who serve as peer reviewers or external evaluators, the survey generally indicated that there are human resource needs to be met with respect to full-time staff to carry out managerial, technical/professional and administrative responsibilities. In this context, while multiple functions are stated in the legislation, not all EQAAs currently have the institutional capacity to deliver services in all areas. Consequently, prioritization of the services to be offered by EQAAs should be focused on areas in which there is an assessed need at the national level. Table 5.2

Table 5.2 Functions of national accreditation bodies established since 2004

Country	Programme approval	Registration	Institutional accreditation	Programme accreditation	Transnational recognition	CSME skills certificate	Title conferment	Equivalence assessment
Antigua and Barbuda	✓	✓	✓	✓	✓	✓	✗	✓
Barbados	✓	✓	✓	✓	✓	✓	✓	✓
Belize	✓	✓	✗	✓	✓	✗	✓	✓
Dominica	✓	✗	✓	✓	✓	✓	✗	✓
Grenada	✓	✗	✓	✓	✓	✗	✗	✓
Guyana	✓	✓	✗	✓	✓	✓	✓	✓
St Vincent and the Grenadines	✓	✓	✓	✓	✓	✓	✓	✓
Suriname	✓	✗	✗	✓	✓	✗	✗	✓
Trinidad and Tobago	✓	✓	✓	✓	✓	✗	✓	✓

provides details of the functions ascribed to national accreditation bodies in the legislation enacted between 2004 and 2011 with the guidance of the CARICOM draft bill.

Unlike table 5.1, Jamaica and St Christopher and Nevis are not included in table 5.2 as the legislation in these two countries predated the draft bill circulated by CARICOM in 2002. No valid comparison can be made between the powers and functions of the EQAAs established since 2004 and the functions enshrined in those pieces of legislation developed in the 1980s and 1990s in a tertiary education environment that was far different from what exists today. The growth of public and private higher education, the expansion of cross-border arrangements and the proliferation of offshore providers are all characteristic of a twenty-first-century higher education environment in which the regulatory function of EQAAs is of critical importance. As stated earlier in this chapter, the legislation in St Christopher and Nevis only addresses the issue of institutional accreditation. In Jamaica, the functions listed in the UCJ Act were specific to the granting of academic awards and distinctions, determining the conditions governing the grant of such awards and approving courses of study to be pursued by candidates for the awards (UCJ Act, section 4). Consistent with good practice for EQAAs globally, the power to grant and confer degrees, diplomas and certificates (UCJ Act, section 5) would be in conflict with UCJ's role as an accreditation body and the legislation that is now over twenty-five years old, is currently being amended to address this and other issues.

The functions presented in table 5.2 are those that EQAAs are empowered by legislation to carry out and not necessarily those currently being performed by the agencies. The validation and approval of new programmes; assessment and recognition of foreign and transnational institutions and the qualifications they award; and programme accreditation are functions common to all EQAAs. Assessment of the equivalence of qualifications is included for all but the one in Suriname. Registration, which ensures that institutions meet minimum standards to operate in the specific territory, is a function of the EQAAs in Antigua and Barbuda, Barbados, Belize, Guyana, St Vincent and the Grenadines, and Trinidad and Tobago. Registration therefore precedes institutional accreditation (it can serve as a condition of eligibility for accreditation), which EQAAs in Antigua and Barbuda, Barbados, Dominica, Grenada, St Vincent and the Grenadines, and Trinidad and Tobago are empowered to award. The granting of skills recognition certificates, which allow CARICOM citizens to travel and work in accordance with the provisions of the CARICOM Skilled Nationals Act, 1997, is also one of the functions of EQAAs, with the exception of those in Belize, Guyana, Suriname, and Trinidad and Tobago. The conferment of protected titles listed in the legislation is a function only of those EQAAs, in Barbados, Belize, Guyana, St Vincent and the Grenadines, and Trinidad and Tobago.

Governance and Autonomy in Decision Making

While all of the EQAAs are state funded, they are governed by boards or councils comprising representatives of various stakeholder groups, including private and public tertiary level institutions, professional associations, labour organizations, the business community, and government ministries and agencies. The board in Trinidad and Tobago also includes two representatives of the general public. The "Guidelines of Good Practice in Quality Assurance" published by the International Network of Quality Assurance Agencies in Higher Education (INQAAHE 2007) state that an EQAA must be independent; have autonomous responsibility for its operations; and its judgements cannot be influenced by third parties. Varying degrees of autonomy are evident in the legislation within CARICOM. The boards of all EQAAs are appointed by the respective government minister, with the exception of the board of the Accreditation Council of Trinidad and Tobago, which is appointed by the president of the republic. With respect to decision making, particularly on accreditation, the power resides solely with the board in the legislation for EQAAs in Barbados, Belize, Dominica, Grenada, Guyana, Jamaica, Trinidad and Tobago, and Suriname. This is consistent with international best practice. In the case of legislation for EQAAs in Antigua and Barbuda, St Kitts and Nevis, and St Vincent and the Grenadines, decision-making power resides with the minister in accordance with a recommendation from the board. In the interest of increasing public confidence in the decisions taken and strengthening the image of these agencies as autonomous bodies, some consideration should be given to amending the legislation in the future so that decision-making authority rests solely with the governing body.

In keeping with good practice for EQAAs, the legislation within CARICOM establishes appeal systems in many territories. Appeals can be directed to the respective minister in Antigua and Barbuda, Barbados, Dominica, Grenada, and St Vincent and the Grenadines. In the legislation for Belize, Trinidad and Tobago, and Suriname the appeal systems differ. In Belize, appeals are directed to a court of law. Appeals in the case of the Accreditation Council of Trinidad and Tobago would be heard by an Appeals Committee appointed by the president of the republic. In Suriname, an appeal would first be directed to the respective minister, but if the applicant is not satisfied with the outcome the appeal can be presented to the president of Suriname. There is no system of appeal in the existing legislation in Guyana, Jamaica or St Christopher and Nevis, and it is envisaged that this will be addressed when future amendments are made.

Accountability and Transparency

All EQAAs in CARICOM, with the exception of the St Christopher and Nevis Accreditation Board, are required by legislation to submit to the respective minister an annual report on work and activities undertaken and audited financial statements, which are laid in parliament. Through this mechanism, the national bodies are held accountable for the use of public funds in the execution of their functions. Additionally, in accordance with international good practice for EQAAs, the legislation in Barbados, Belize, and Trinidad and Tobago compel the agency to undergo a review of its operations. In Barbados, the minister has the power to "undertake a review of the Council every 3 years to determine its effectiveness and efficiency" (section 25, Act No. 11 of 2004). In the legislation in Belize and Trinidad and Tobago, the cycle for review is six and three years, respectively, and the acts state that the review shall be undertaken in collaboration with regional accrediting bodies and other competent authorities within CARICOM. An assessment of the impact of the operations of the council on the society is also included in the review. These measures contribute not only to ensuring accountability and transparency but impact positively on the integrity of the operations of EQAAs as the bodies are subject to external scrutiny. No agency has so far undertaken a review in accordance with the act.

CARICOM Accreditation Agency for Education and Training

There is before CARICOM member states an intergovernmental agreement to establish the CARICOM Accreditation Agency for Education and Training (CARICOM Secretariat, June 21, 2007). There has been no implementation of this decision as so far the process of securing signatories to the agreement has been a protracted one. The issue of funding for the agency has reportedly been one of the obstacles to its establishment. It is envisaged that this regional agency would, among other things, participate in the external review of national EQAAs; seek to harmonize standards of quality and quality assurance procedures; develop guidelines for good practice; and facilitate mutual recognition agreements. This agency, when operational, would strengthen the regional profile with regard to the integrity of the national and regional systems for external quality assurance. While the regional body may facilitate and co-ordinate arrangements for the accreditation of institutions particularly in territories in which national bodies have not been established (or in the case of a multi-campus regional university such as the University of the West Indies), it was decided that the body itself will have no power to accredit as this power should reside with the national body in the member state in which the educational institution operates. Additionally, direct involvement of the regional body in the accreditation of institutions would compromise the role it is required to play as an oversight body to monitor and evaluate the quality and integrity of the operations of national EQAAs.

It should be noted that there are three regional bodies for specialized accreditation in the fields of medicine and other health professions and engineering. They are the Caribbean Accreditation Authority for Education in Medicine and other Health

Professions; the Caribbean Accreditation Council for Engineering and Technology; and the Greater Caribbean Regional Engineering Accreditation System. While programme accreditation is listed among the functions of all EQAAs in CARICOM, there is consensus among the bodies that in fields requiring specialized accreditation this should be undertaken by the relevant regional accrediting bodies. The accreditation awarded by these specialized agencies is recognized by the national EQAAs in CARICOM. There are other initiatives being pursued within CARICOM, such as the development of the CARICOM qualifications framework (this is a regional framework which will be supported by national qualifications frameworks in member states), which would strengthen mechanisms for determining the equivalence of qualifications and comparability of standards for tertiary education across the region. Together these developments would serve to assure and improve quality and facilitate increased mobility of skilled persons throughout the CSME.

Conclusion

For over a decade, governments of CARICOM countries have steadily sought to increase the investment in human development and, in particular, to commit more resources to the expansion of tertiary education. The value of higher education to the development of the region is evident in initiatives such as the establishment of the Council for Human and Social Development which, subject to the provisions of Article 12 (*Functions and Powers of the Conference*) of the Revised Treaty of Chaguaramas, is responsible for the promotion of human and social development in the Community. One of the major responsibilities of the Council for Human and Social Development is to promote the development of education through the efficient organization of educational and training facilities in the Community.

It is widely recognized that education, as an investment, has high potential rates of individual and social return, and that the majority of jobs in the new knowledge-driven economy will require post-secondary education. Additionally, human development is so central to economic growth; national security and social cohesiveness that the peace, prosperity and security of individual nations are inextricably linked to the efficiency and effectiveness of their higher education systems.

Future development of the EQAAs in CARICOM member states is contingent upon the financial and human resources available to these embryonic bodies. Since they are funded by the state, this places an increased burden on the governments of the region and, ultimately, the taxpayers. The autonomous (or semi-autonomous), publicly funded corporate bodies, which have evolved out of the CARICOM draft legislation, should seek to reduce reliance on government subventions by generating revenue from fees for services and from events such as conferences and workshops. EQAAs must also maintain sufficient professional and support staff to ensure that higher education institutions and other stakeholders can access services and support in an efficient and effective manner. A significant investment in professional development and training is required to strengthen institutional capacity in the EQAAs. Failure to do so can undermine the integrity of the system and negatively impact the sustainability of quality assurance initiatives, ultimately leading to a loss of confidence in the EQAA and in the quality assurance system in general.

This study makes a significant contribution to the knowledge based on national quality assurance systems in the CARICOM. As a region that is distinct in many respects, CARICOM has given birth to a type of hybrid EQAA with a multiplicity of roles and functions. While there are many challenges inherent in this model, some of the features of the emergent EQAAs can potentially be adapted in other small states in which the scarcity of resources increases the value of approaches through which countries can derive economies of scale.

Acknowledgements

The author acknowledges the contribution of the chief executive officer or chairperson of the following bodies to the survey: National Accreditation Board of Antigua and Barbuda; Barbados Accreditation Council; National Accreditation Board of Dominica; National Accreditation Council of Guyana; University Council of Jamaica; St Christopher and Nevis Accreditation Board; National Accreditation Board of St Vincent and the Grenadines; National Council for Accreditation Suriname; Accreditation Council of Trinidad and Tobago. The author also wishes to thank the chairperson of National Council for Accreditation Suriname for providing an English translation of the act.

References

Antigua and Barbuda. 2006. The Accreditation Act 2006. No. 4 of 2006 (Antigua and Barbuda).
Barbados. 2004. Barbados Accreditation Council Act 2004–11 (Barbados).
Belize. 2004. National Accreditation Council of Belize Act 2004. No. 20 of 2004 (Belize).
CARICOM Secretariat. 1973. "Treaty Establishing the Caribbean Community (1973)". http://www.caricom.org/jsp/community/original_treaty-text.pdf
———. 2001. "Revised Treaty of Chaguaramas Establishing the Caribbean Community Including the Caribbean Single Market and Economy". http://www.caricom.org/jsp/community/revised_treaty-text.pdf.
———. 2002. "The National Accreditation Council [Agency] Bill No. 2002". CARICOM Secretariat, Georgetown, Guyana.
———. 2007. "Inter-Governmental Agreement to Establish the CARICOM Accreditation Agency for Education and Training". Unpublished document. CARICOM Secretariat, Georgetown, Guyana. June 21.

Commonwealth of Dominica. 2006. Accreditation Act 2006. No. 13 of 2006 (Dominica).
Girvan, Norman. 2007. "Towards a Single Development Vision and the Role of the Single Economy". Paper prepared in collaboration with the CARICOM Secretariat and the Special Task Force on the Single Economy. Twenty-Eighth Meeting of the Conference of Heads of Government of the Caribbean Community (CARICOM), July 1–4, 2007, Needham's Point, Barbados. http://www.caricom.org/jsp/single_market/single_economy_girvan.pdf.
Grenada. 2011. Accreditation Act 2011. No. 15 of 2011 (Grenada).
Guyana. 2004. National Accreditation Council Act 2004. Act No. 12 of 2004 (Guyana).
INQAAHE (International Network of Quality Assurance Agencies in Higher Education). 2007. "Guidelines of Good Practice in Quality Assurance". http://www.inqaahe.org/admin/files/assets/subsites/1/documenten/1231430767_inqaahe---guidelines-of-good-practice%5B1%5D.pdf.
Jamaica. 1987. The University Council of Jamaica Act 1987. Act 23 of 1987 (Jamaica).
———. 1991. The University Council of Jamaica (Amendment) Act 1991. Act 16 of 1991 (Jamaica).
Laws of Trinidad and Tobago. 2004. Accreditation Council of Trinidad and Tobago Act, Chapter 39:06. Act No. 16 of 2004 (Republic of Trinidad and Tobago).
———. 2007. Accreditation Council of Trinidad and Tobago (Amendment) Act, Chapter 39:06. Act No. 16 of 2007 (Republic of Trinidad and Tobago).
———. 2008. Accreditation Council of Trinidad and Tobago (Amendment) Act, Chapter 39:06. Act No. 10 of 2008 (Republic of Trinidad and Tobago).
Republic of Suriname. 2007. NOVA. Act 74 of 2007 (Republic of Suriname).
Republic of Trinidad and Tobago. 1971. Cabinet Minute 1297 of December 9, 1971. Unpublished document, Republic of Trinidad and Tobago.
Saint Christopher and Nevis. 1999. The Saint Christopher and Nevis Accreditation of Institutions Act 1999. Act No. 21 of 1999 (Saint Christopher and Nevis).
———. 2001. The Saint Christopher and Nevis Accreditation of Institutions (Amendment) Act 2001. Act No. 9 of 2001 (Saint Christopher and Nevis).
Saint Vincent and the Grenadines. 2006. Further and Higher Education (Accreditation) Act 2006. Act No. 35 of 2006 (Saint Vincent and the Grenadines).
Wickham, P.W., C.L.A. Wharton, D.A. Marshall and H.A. Darlington-Weekes. 2004. "Freedom of movement: The cornerstone of the Caribbean Single Market and Economy (CSME)". Paper prepared for the Caribbean Policy Development Centre (CPDC). http://sta.uwi.edu/salises/workshop/papers/pwickham.pdf.

6 | Ethics and Quality Assurance
Purpose, Values and Principles

ANNA KASAFI PERKINS

The three teachers contended in their motion that the ministry ought not to conduct its own secret assessment of their qualifications, but instead abide by the ACTT's accreditation.
— Azard Ali, *Trinidad and Tobago Newsday* (online)

It's a real scam going on out there and Jamaican students are getting caught.
— Dr Ethley London, then-executive director of UCJ speaking on bogus degrees, *Jamaica Observer*, May 4, 2007

Quality assurance (QA) in higher education is a process of evaluating, ensuring and enhancing standards of educational provision. QA usually entails two aspects – internal (IQA) and external processes (EQA). Indeed, QA begins as an internal process that is normally monitored or assessed through an external quality assurance agency (EQAA) (National Assessment and Accreditation Council India and Commonwealth of Learning 2006).

Any process of assessing performance in relation to standards assumes much ethical content, that is, asserting and living out values such as commitment to truth-telling, promise keeping, excellence (quality), transparency, accountability, integrity and responsibility. (Furthermore, concerns with goodness, rightness, fairness, virtue, fitness and purpose aver to ethical categories.) Such an understanding of quality based on judgement and standards contains an ethical "test". Higher education institutions make promises by virtue of their mission and vision and the extent to which these "promises" are kept has come increasingly under scrutiny, especially in a context where there are growing demands for "value for money" and "greater efficiency" (Beckles, Perry and Whiteley 2002).

Quality cannot exist without ethics: "there is no policy, no behaviour, no practice, no value that does not have a direct impact on the condition of quality on a college or university campus" (Bogue and Saunders 1992, 258). That being the case, QA functions to ensure and enhance ethics in academic and institutional processes. Indeed, it is oftentimes during internal quality audits or external quality assurance (EQA) processes like accreditation that evidence of academic misconduct or institutional lapses become evident. Internal quality assessments may reveal padded enrolment figures, publication fraud, cheating on examinations and lecturer absence from classes, to name a few. Accreditation processes have uncovered, for example, the presence of "a culture of fear", absentee governing boards, flawed management and even sexual harassment of students and employees (Bogue and Saunders 1992; Fain 2013a). Even in the face of such revelations, the ethical import of QA in higher education is often unacknowledged as there is little scholarship on the relationship between the two (see Margetson 1997; Nayebpour and Koehn 2003).

In the Commonwealth Caribbean, the importance of ensuring ethics in academic life is reflected in the standards of national accrediting bodies like the University Council of Jamaica (UCJ) with its focus on integrity (Standard 10), as well as the Barbados Accreditation Council (BAC) and the Accreditation Council of Trinidad and Tobago (ACTT), for which both ethical leadership and financial integrity are criteria for institutional accreditation. Similarly, the regional Caribbean Accreditation Authority for Education in Medicine and other Health Professions (CAAM-HP), while not outlining a specific criterion dealing with ethics, covers ethical practice, education and formation extensively throughout its standards (ED 10, 20; MS 7–8, 25–31) (see appendix 6.1). Similarly, the Caribbean Accreditation Council for Engineering and Technology (CACET) in their 2008 Declaration links quality and ethics as it promises that the Council will be "establish[ed] on the principles of advancing the quality of accredited programmes; demonstrating accountability; encouraging continuous self-assessment and improvement; employing fair, equitable, transparent and publicly announced procedures; respecting variety and diversity in higher education; honouring the autonomy and academic freedom of programmes and their professoriate; and maintaining economical and minimally intrusive methodologies and practices" (CACET 2008, 1).

This chapter provides QA practitioners with an introduction to the role/purpose of QA in ensuring ethics in higher education. It does so by exploring the commitment related to ethics espoused by selected Caribbean accreditation agencies – UCJ, CAAM-HP, BAC, CACET and ACTT – in their core values and mission. At the same time, it clarifies the notion that QA itself is expected to be an ethical practice and is undergirded by the same values it ensures. In so doing, the chapter discusses the kinds of ethical lapses that have taken place in the process of accreditation; these practices have been seen to be on the increase. It takes account of examples of alleged or potential ethical lapses or misconduct of three EQAAs, two of which are drawn from the Caribbean region. It closes by outlining briefly an ethical framework for EQAAs that starts with values identification and ends with a process of self-evaluation. Perhaps this skeletal framework, which can be fleshed out by individual EQAAs, may contribute to diminishing ethical infringements in the QA process. The focus on EQAAs is by no means a suggestion that EQAAs are more susceptible to ethical lapses, but more a function of the fact of the public nature of the EQAA process and reports. Internal quality assurance processes often generate reports for institutional improvement and are therefore not made public. No higher education institution in the Caribbean region presently makes public its internal quality assurance reports.

Ethical Purpose of Quality Assurance

Accreditation, which is probably the most widely known and respected form of QA (Bogue and Saunders 1992), refers to the status granted "through a socially legitimized agency [that] provides public assurance on the ability of higher education institutions or their programmes to fulfil stated purposes and compliance with previously determined quality criteria" (Scott 2010, 28); it is also the process of assessing that attainment (Beckles, Perry and Whiteley 2002). Academic programme review is another kind of QA mechanism and Bogue and Saunders (1992) detail the ethical purpose of such internal reviews: "To ascertain whether the program is being operated according to *ethical standards* set forth by the institution and/or the governing body, whether the *purposes* of the program are commensurate with the *purposes* of the institution as a whole, or whether the operations of the institution or a particular program are characterized by *good* management practices, efficiency, and *integrity*" (Bogue and Saunders 1992, 139; emphasis added).

Undoubtedly, the same ethical purpose is at play in programme and institutional accreditation. Bogue and Saunders assume the presence of an ethical framework within a higher education institution that can serve as a basis for evaluating academic programmes, and, by extension, the institution itself. This is clearly the perspective of UCJ, BAC and ACTT, which specifically assess the presence of such a framework by way of standards (see appendix 6.1). In order to do so, however, the agencies begin with their own mandate and ethical framework. In keeping with international best practices all of these criteria are publicly available. (This is not the case for all the EQAAs in the region.) They claim ethics as a central part of their own way of being. This is clearly the basis upon which they can demand ethical practices and systems from the academic institutions that they assess. These accreditation agencies elaborate the importance of ethics in framing their core values. BAC, for example, lists five values which are at the core of its own mission. Included among these is integrity (2012, 5). BAC also promises that it will place "fairness, honesty, objectivity, accountability and transparency" at the forefront of its own policies and practices as well as in the implementation of accreditation and related processes in order to establish integrity *as central to educational quality* (2012, 5). ACTT lists its core values as: accountability, commitment to personal growth and development, customer focus, integrity, and teamwork, and trust. UCJ spells out carefully its core values, among which are: *integrity and ethics* – objectivity, honesty, the establishment of trust and confidentiality while consistently operating in an ethical manner; *transparency* – openness and fairness in operations; *respect* – values the diversity of our stakeholders. Unlike BAC or ACTT, however, as demonstrated by the previous sentence, UCJ outlines what is meant by each value. ACTT, on the other hand, makes the claim that "[it] will be guided by its core values which outline the types of behaviours and attitudes that will be crucial for its success" (actt.org.tt) without detailing the consensus around the meaning of these values.

It is important to be able, not only to list, but also define the values that undergird practice otherwise there will be a lack of clarity regarding what the values mean and they will therefore be difficult, if not impossible, to attain. Furthermore, in a diverse social space it is necessary to articulate the content of such values rather than assume generally accepted meanings. CAAM-HP and CACET currently lack defined core values or an explicit ethical framework although it is possible to ferret these out by analysing their descriptions of their accreditation systems. CACET is clear that it "is committed to *fair, equitable* and *transparent* processes and to mechanisms that lead to continued improvement in the operations of CACET and the programs it accredits" (CACET Accreditation Manual 2009, 2; emphasis added). These are the same values espoused by other EQAAs.

Similarly, CAAM-HP states that the accreditation system used by the CAAM-HP is based on that of the Liaison Committee on Medical Education. This system functions with a clear, authoritative mandate, operates *independently* of governments and institutions, uses preset *standards*, draws on external reviewers and involves *self-evaluation* and site visits (caam-hp.org; emphasis added).

There are values that are implicit in this CAAM-HP description: independence, transparency, and accountability of and to stakeholders – the very values made explicit in the core values and operational mandate of CAAM-HP's counterpart national EQAAs – UCJ, BAC and ACTT. It is important that CAAM-HP and CACET make their values explicit; similarly, as will be discussed below, CAAM-HP should carefully articulate criteria on ethics to encompass the various standards already being assessed. There is significant overlap among the core values of the three national accrediting bodies; the value present among all three is integrity. Integrity is a key value as it speaks to the commitment to living out the values espoused. It also appears to be a cipher for the word ethics, as discussions like that of Bogue and Saunders above suggest.

Undermining the QA Ethics Mandate

Such values are undermined significantly by unethical practice in external quality assurance globally. Accreditation processes are subject to unethical practices ranging from fraud and corruption to professional back-scratching or falsely claiming credentials. In fact, Hallak and Poisson (2007) argue that accreditation and certification processes worldwide are increasingly undermined by fraud. Referencing a study undertaken in the Ukraine, Hallak and Poisson reveal that few successful accreditation applications in that nation were not the result of some form of bribery. Among the forms of corruption they identified are payment of bribes, circumvention of accreditation procedures by higher education providers through franchising schemes, accreditation processes based on non-transparent criteria, the rise of accreditation mills to support the efforts of diploma mills, and non-accredited institutions falsely issuing accredited degrees. Fraud is not the only kind of unethical practice that undermines accreditation processes; conflicts of interest, ignoring internal rules and regulations, and stakeholder disenfranchisement are also relevant, as the brief cases below demonstrate.

Conflicts of Interest

The potential for a conflict of interest arises when one's duty to make decisions in the public's interest is compromised by competing interests of a professional, personal or private nature, including but not limited to pecuniary interests (Council on Education for Public Health, 2009). The evaluative nature of the QA process and its significant implications for reputation, national development, personal success, funding, and so forth, demands the independence and competence of the review teams and the transparency and effectiveness of the evaluation process (see Perkins and Dallas 2013).

Accrediting Commission for Community and Junior Colleges and City College of San Francisco

A well-known case concerning alleged misconduct by an accrediting agency comes from the United States. The Accrediting Commission for Community and Junior Colleges (ACCJC), part of the Western Association of Schools and Colleges, has been accused of acting unethically when it slapped City College of San Francisco (CCSF) with a severe sanction and then voted to strip the college of its accreditation. The California Federation of Teachers, whose membership includes the local union representing faculty members of CCSF, filed a complaint alleging that "the accreditor's review of City College was deeply flawed, tainted by conflicts of interest and in violation of [its own rules and] state and federal laws" (Fain 2013b). The union further alleges that the ACCJC lacked a policy on the composition of site visit teams. The claim is that the teams that visited CCSF "were heavy on administrators and light on professors" (Fain 2013c). They also detail conflicts of interest in the review centring around the presence of the ACCJC president's husband, himself an administrator at a California community college, on one of the CCSF's site visit teams.

Interestingly, among the deficiencies identified by the commission at CCSF is that the college failed to fix a set of problems they had been required to address from as early as 2006, including failure to track student outcomes and operating with three days' cash reserves (Fain 2013c). CCSF is disputing that the recommendations were mandatory as they were "quality improvements" so "the commission cannot legally enforce its suggestions with the same vigour as its standards for accreditation eligibility" (Fain 2013b). The general dissatisfaction of faculty unions and some administrators of community colleges in Western California with the accreditor is clear in the description of the commission as "overly antagonistic", accompanied by calls for the Education Department not to renew the commission's federal recognition. The dispute with CCFS is neither the first nor ACCJC's only.

The Education Department has since sanctioned the accreditor stating that it is not in compliance with several federal regulations and was liable for sanctions unless the problems were corrected within twelve months (Kelderman 2013). Among the areas in which the commission has been sanctioned are the policy on the composition of visiting accreditation teams and policy on conflicts of interest. "The participation of the spouse of the president of the ACCJC on an evaluation team has the appearance to the public of creating a conflict of interest ... an appearance of bias of the commission in favour of the team's position over that of the institution" (Kay Gilcher, director of the Department of Education's Accreditation Group, cited in Fain 2013c). The commission is also required to provide a clear explanation to institutions concerning areas where they fall below standard along with a time frame for improvements. Interestingly, the CCSF has since appealed the commission's sanction but without any reference to this ruling by the education department as part of an argument in appealing the ACCJC's sanction. An overturn of the accreditor's decision will only happen if CCSF can prove ACCJC made procedural errors in its decision to remove the college's accreditation. The City of San Francisco itself recently entered the fray in announcing that it would be suing the ACCJC, charging it with "using inappropriate measures in evaluating

City College of San Francisco... The suit alleges among other things, that the commission was punishing the college for its opposition to budget cuts at a time that commission leaders agreed with state officials that cuts were appropriate" (Jaschik 2013). ACCJC, of course, continues to maintain that is did not treat CCSF unfairly. The impact of the city's decision to sue is unclear; however, as the accreditors are not managed by the state and accreditation is a voluntary process although having that status does affect state support and federal financial aid programmes.

Jamaica Tertiary Education and Research Commission, University Council of Jamaica and Council of Community Colleges of Jamaica

The potential for other kinds of conflicts of interest or role confusion was present in the case of the newly formed tertiary regulator, Jamaica Tertiary Education and Research Commission (JTEC), the UCJ and the Council of Community Colleges of Jamaica (CCCJ). Jamaica was expected to pass legislation by the end of 2013 to bring on board a new regulatory body that is billed to "supersede" other entities such as UCJ, CCCJ and the Tertiary Unit of the Ministry of Education (*Jamaica Observer*, January 26, 2011). This has not yet happened. The reduction of duplication of responsibilities and the need to "address lack of resources and resulting constraints in programme implementation experienced by some of the organisations that JTEC replaces" (*Jamaica Observer*, January 26, 2011) are presented as imperatives for the formation of a regulator. As the mandate for the JTEC was elaborated, there was the need to clarify its relationship with UCJ. Questions that emerged in light of the original Cabinet decision to make UCJ a subcommittee of JTEC included (Henry-Wilson, email, August 19, 2013): How many of the current functions of the UCJ would JTEC assume? How would the independence of the accrediting agency be maintained? These questions arise in the face of JTEC being expected to register tertiary institutions, which is currently part of the UCJ mandate. Importantly, how will the two entities relate in terms of reporting relationship? As the case of the ACCJC demonstrates, accrediting bodies need themselves to be subject to some form of oversight and "accreditation". Will JTEC provide that? The International Network of Quality Assurance Agencies in Higher Education's Guidelines for Good Practice (2007) details the necessity of internal self-review and external reviews at regular intervals as being good practice for EQAAs like UCJ. However, ambiguous reportage like "Although the University Council of Jamaica (UCJ) provides accreditation for tertiary programmes following the graduation of a cohort, Henry-Wilson (that is, the executive director of JTEC), pointed out that it would be the function of the JTEC to register institutions at the time they are established to guarantee quality assurance at the outset" (Wilson 2013) and the lack of publicly available information on the regulator, for example, a dedicated website, deepens the confusion. The role of JTEC in undertaking accreditation is unclear in that statement, as in this one: "The University Council of Jamaica provides accreditation for tertiary programmes but the University of the West Indies is the only one where the institution is fully accredited. The JTEC will attempt to get other institutions to do the same, while the UCJ continues to assess individual programmes" (*On the Ground News Report*, https://www.facebook.com/onthegroundnews/posts/636345983059445). Will institutional accreditation move to JTEC?

Favourably, as the work of developing the JTEC has progressed, these potential conflicts have been addressed. As the executive director of JTEC informs, currently registration with the UCJ is a precursor to an application for accreditation. However, as things presently stand in the tertiary sector in Jamaica, because registration is voluntary "there are several institutions operating in our tertiary education space that are not registered. This applies to local, offshore and hybrid institutions" (Henry-Wilson, email, August 19, 2013). JTEC will therefore have responsibility for the compulsory registration of all higher education institutions in Jamaica: "This will be J-TEC's responsibility and, further to ensure there are minimum quality assurance standards for each institution primarily as these relate to governance, financing and infrastructure for programme offerings. One of the requirements for licensing by J-TEC will be proof of application to UCJ for accreditation of programmes being offered by institutions. Failure to obtain accreditation within a designated period will result in cancelling of licence" (ibid.).

The provision of licences to operate as a tertiary level institution is clearly now part of the mandate of JTEC; such licensing to provide higher education was not previously part of the landscape in Jamaica. Now licensing is being carefully tied to matters of QA, both internal (governance, programme offerings, financial management) and external (registration and proof of application for programme accreditation). The conflicts addressed should flag for JTEC the necessity of articulating its own ethical framework, including its core values, ethical criteria for performance of its role and for assessing institutions for registration (similar perhaps to the Accreditation Standards of UCJ, BAC and ACTT). The matter of the external accreditation of UCJ still requires attention, also.

On another matter, the UCJ's mandate to accredit as well as award degrees appears to be a further conflict of interest between two distinct functions. If the UCJ Act gives it the power to award degrees and it is the only accrediting body, will it accredit its own degrees? This potential conflict is highlighted by an earlier iteration of the proposed merger between UCJ and JTEC, as was pointed out by Alfred Sangster, former president of the University of Technology, in a letter to the editor of the Jamaica *Gleaner* ("An Academic No-no", May 1, 2013). Sangster is concerned about the proposed merger of the UCJ with the CCCJ under the JTEC umbrella to "enable greater efficiency and effectiveness". Such a merger raises, in his mind, questions of authority and integrity. Since the UCJ accredits programmes of the tertiary sector, including those of the CCCJ, the proposal to merge an institution linked with community colleges with one accrediting programmes at university level may lead to "the lower level accrediting the higher level! It's an academic no-no!" (Sangster 2013). However, it seems Sangster's concern emerged from a misunderstanding on his part of an earlier proposal It was actually proposed that CCCJ be abolished and JTEC assume CCCJ's mandate (Henry-Wilson, email, August 19, 2013). Clearly, the lack of clarity of the unfolding mandate of the JTEC contributed to the misunderstanding. It also highlights the importance of transparency in ensuring ethical practice. Many kinds of unethical activity happen in the grey areas where matters are unclear.

Promise Keeping and Stakeholder Involvement

From Antigua comes a case concerning the accreditation status of American University Antigua (AUA), a private for-profit medical school in Antigua and Barbuda, which began operations in 2003 under government charter. In 2011, alarm was raised in the media about the accreditation status of the school because it was reported that graduates from the medical programme were not licensed to practice in Antigua and Barbuda, since the medical school was not accredited ("Standing by Our 'Accreditation' Story", *Antigua Observer* [online], March 7, 2011). The question of AUA's accreditation status and the licensing of its graduates was raised as far back as 2009. In that year, AUA filed a lawsuit directly suing the state medical board in Arkansas for putting the school on its list of "disapproved" schools, that is, medical schools deemed unacceptable; AUA graduates therefore could not qualify for licences in Arkansas (Lederman 2009). AUA alleged that the Arkansas medical board was in collusion with the national medical groups to restrict the approval of foreign medical schools and their graduates "under the guise of protecting the quality of the practice of medicine in the United States" (Lederman 2009). (The outcome of that lawsuit is unclear.)

In the wake of the 2011 allegations, the Antigua and Barbuda National Accreditation Board (ABNAB) responded to say that the AUA's 2003 Charter issued by the government was in order and that the board had accredited the AUA School of Nursing and was in the process of registering the School of Medicine and the Vet School. Importantly, the government maintained that AUA had its full confidence: "The government of Antigua and Barbuda stands by the American University of Antigua, which is registered and licensed by the State of New York" ("Standing by Our 'Accreditation' Story", *Antigua Observer* [online], March 7, 2011). This claim to registration and licensing is not to be mistaken for accreditation (although it clearly has been even by the Antigua government). The Antigua and Barbuda government in their official statement said: "By the Charter, the Government has recognized and continues to recognize the New York State Board as a duly authorized accrediting body which has accredited the American University of Antigua, College of Medicine" (Government of Antigua and Barbuda 2011). Into the mix was added, the fact that the ABNAB was started five years after the AUA had come into being. At the time the story broke, the ABNAB was in the process of registering institutions and had not yet accredited any. Some of the public commentary on the situation turned on the ABNAB, which was accused of being "dysfunctional and not able to get its act together" as well as the government for having failed to do due diligence before offering AUA a charter ("Standing by Our 'Accreditation' Story", *Antigua Observer* [online], March 7, 2011). The question of the role of a regional accreditation agency was also thrown into the ring.

The Antigua and Barbuda Nursing Council entered the dispute expressing displeasure with the National Accreditation Board. "The council would like to air its utter disgust with the Antigua and Barbuda National Accreditation Board (ABNAB) for totally ignoring the professional body that is responsible for regulating nursing as well as the nursing profession as a discipline by not involving it in the registration process, especially as it relates to the training of nurses" ("Nurses Council Not Happy with Accreditation Board", *Antigua Observer*, March 5, 2011). The council charged that the ABNAB violated its own regulations, which stipulate that it will consult with various professional disciplines in decision making. The AUA suspended applications to the School of Nursing once it failed to receive approval from the nursing council and its subsequent rejection by the Commission on Graduates of Foreign Nursing Schools in August 2011. The nursing school was eventually closed. The vet school was closed in December 2011 and students transferred to St George's University School of Veterinary Medicine in Grenada. Among other matters, this case seemed to be about concern and confusion between institutional and programme accreditation, licensing as well as registration. Approval by the New York State Education Department and recognition by the Medical Board of California allow students to obtain residency positions and secure clinical clerkships in those states but this does not accredit programmes. The president of the university may have overstated their position therefore when he stated during the row, "We have accreditation. We are fully approved and our graduates are eligible for licensure" (Holder 2011). Happily, perhaps in keeping with further claims by the president that "we welcome accreditation and we welcome additional review" (Holder 2011), the AUA approached CAAM-HP for accreditation. In 2012, CAAM-HP granted provisional accreditation until 2014 to the AUA medical programme (see assessed programme report on http://caam-hp.org/assessedprogrammes.html).

The ABNAB case calls up questions concerning the relationship between EQAAs and other stakeholder groups, like the government and the powerful higher education institutions. EQAAs in small island developing states are often subject to political pressures, as Extavour and Whittington (2011) discuss. The political directorate in many small societies can apply pressure because of size, legislative mandate, dependence for funding and operational support from the state. As in the ABNAB case, the whispered concern was the fear of the impact of the loss of reputation for AUA on the Antiguan and Barbudan economy; the AUA has contributed significantly to the local economy. Nonetheless, Caribbean QA professionals Extavour and Whittington argue that through strategic management, and a project-based approach for each stakeholder institution, the EQAAs "can initiate, develop and maintain a high degree of 'independence' in an environment that is increasingly being driven by economic considerations, within a social context of public and private higher education institutions" (2011, 3).

Ethical Framework

Quality in higher education exists in a reciprocal relationship with ethics: there is no quality without ethics and ethical practice redounds to educational quality. This has been recognized by some EQAAs in the Commonwealth Caribbean, as the discussion above has detailed. This relationship is captured, for example, in BAC identifying one of the defining characteristics of quality in institutions is "a culture which embraces integrity and ethical conduct" (2012, 6). The centrality of ethics to quality in higher education, calls for a QA system that is itself framed around ethics – values, principles and process. UCJ makes this clear when it claims in its mission to intend to "increase the availability of tertiary level training in Jamaica through a robust quality assurance system that ensures excellence, transparency, integrity, and adherence to standards" (http://www.ucj.org.jm/content/mission-vision-statement). UCJ recognizes the importance of certain values in a national QA system; included among these are, of course, integrity and transparency, values inherent in the process of QA.

Table 6.1 takes account of the work of CACET, CAAM-HP, BAC, ACTT and UCJ, and the learning points from the case studies to outline a brief framework for QAAs to infuse ethics into their process and practice, beginning with the identification of core values.

This framework can also be adapted by internal and external quality assurance agencies with mandates akin to that of JTEC. The framework assumes the acceptance of the ethical mandate of QA, as has been argued for in this chapter. With this acceptance, it is necessary to identify the values that underlie the ethics mandate of QA in higher education. A good place to begin is with fitness for purpose, the model of QA widely used in higher education institutions across the world. The implied concern with "fitness of purpose" can then be surfaced, as the worthwhileness of the purpose cannot be assumed but must be argued for and demonstrated. The mission and core

Table 6.1. An ethical framework for quality assurance

An ethical framework for quality assurance: Putting values into action for quality
Identify values – Based on the underlying values identified in the quality assurance process, including independence, quality, transparency, accountability, integrity, legitimacy and credibility. Take careful account of context and limit number of values.
Define values – Individual explanations that clearly distinguish one value from the other.
Describe standards of performances – Describe how these values will be reflected in the work of the external quality assurance agency, for example, truth-telling is reflected in accurate reporting and clear guidelines.
Outline policy – Related to relevant values, for example, policy on conflicts of interest; composition of team; relationship to stakeholders, etc. These should be made public.
Establish self-evaluation process – Intent is to assess the ethical purpose, values and performance of the EQAA. Include external reference points, such as judgement of knowledgeable peers, international codes of ethics or relevant legislation or policies. The evaluation process itself should reflect the values of the EQAA (independence, autonomy, fairness, transparency and integrity).

values of the agency and higher education institution must then be subject to scrutiny. How to determine whether the purpose is actually "fit" will require a process that is transparent, accountable, independent and dedicated to quality (excellence). This has been recognized by many QA agencies and so the values articulated may simply be adopted and adapted. It is important, however, not have too many core values.

Whichever core values are identified as being the centre of QA, it is necessary and important to define these, so that consensus about meaning will contribute to processes that are truly reflective of these core values. Integrity as a core value that seems to be important to all of the EQAAs explored, in particular, needs concise definition as it seems to be both a value and a cipher for ethics. This can be source of confusion and reduce the potential for the term to be beneficial. Each of the core values can then be linked to various processes and procedures through which the agency carries out is mandate. So the value of truth-telling will be reflected in accurate reporting on findings and well-written guidelines for report preparation; similarly, independence should be linked to the selection and training of evaluators.

Once this has been done, it is necessary to capture these performance standards in policy guidelines to cover the major areas of the EQAA's functioning, for example, conflicts of interest, confidentiality of information and relationships to stakeholders. In keeping with the values like truth-telling and transparency, these policies should be made public. This will ensure that the "ethical test" of the ethics mandate of the

EQAA can be undertaken by its various publics. The feedback from such stakeholders should factor into a clearly defined self-evaluation process through which the EQAA will assess its ethical purpose, values and performance. Other external reference points such as international codes of ethics or relevant legislation, for example, the act establishing the agency. The findings of such self-evaluations can feed into the regional QA mechanisms for EQAA if and when this is established. Similarly, such findings would be important when the EQAA participates in International Network of Quality Assurance Agencies in Higher Education's voluntary reviews of compliance with its guidelines for good practice.

The ethics mandate of QA may not be recognized by many in higher education, including quality professionals. The potential harm from ethical wrongdoing in EQAAs, as detailed briefly in the cases above, is a jarring reminder of the importance of revisiting and recommitting to ethics in QA in higher education. The skeleton of an ethical framework for QA, detailed above, provides external agencies and internal units having QA mandates with criteria to ensure and evaluate the ethics of their process and outcomes.

Appendix 6.1. Accreditation standards dealing with ethics: ACTT, BAC, CAAM-HP, CACET and UCJ

Agency	Standard
Accreditation Council of Trinidad and Tobago	Criterion: The institution's system of governance ensures ethical decision making and efficient provision of human, physical and financial resources to effectively accomplish its educational and other purposes.
	Standard 2.1: The institution's governance and administrative structures and practices promote effective and ethical leadership that is congruent with the mission and objective of the institution.
	Standard 2.3: The institution has sound financial policies and capacity to sustain and ensure the integrity and continuity of the programme offered at the institution.
Barbados Accreditation Council	Criterion: The institution's system of governance ensures ethical decision making and efficient provision of human, physical and financial resources to effectively accomplish its educational and other purposes.
	Standard 2.1: The institution's governance and administrative structures and practices promote effective and ethical leadership that is congruent with the mission and objective of the institution.
	Standard 2.3: The institution has sound financial policies and capacity to sustain and ensure the integrity and continuity of the programme offered at the institution.

Caribbean Accreditation Authority for Education in Medicine and other Health Professions (CAAM-HP)	While not detailing a specific standard for ethics or integrity, CAAM-HP demonstrates a commitment to institutional and programme ethics in several ways.

In Part 2 of its Standards (Explanatory Annotations of its Accreditation Standards), it clarifies the operational meaning of standards set out in narrative form in Part 1. In Section 1: Institutional Settings, Governance and Administration, CAAM-HP details that the governing body overseeing the medical school should be composed of individuals who "have no clear conflict of interest in the operation of the school, its associated hospitals, or other related teaching or service facilities" (CAAM-HP, 3; IS 3, 14). Care is to be taken to select students who possess the intelligence, integrity, personal and emotional characteristics necessary to make them become effective physicians (MS-5). This has to be undertaken through a selection process that is transparent, having established criteria for admission and non-discriminatory (gender, racial, sexual orientation, cultural and economic diversity, disabled) (MS-7 & 8).

Similarly, the learning environment should demonstrate this non-discrimination and should have clear and public policies, standards and procedures regarding evaluation, advancement and graduation and disciplinary action (MS-25–31). Specifically, the doctors graduated should be knowledgeable of international codes of conduct for health professionals and practice within the laws and codes of conduct of the profession for the country in which they are employed (ED-10). To this end, the medical school is required to teach medical ethics (ED-20); ethical conduct is a required and essential part of clinical instruction. "The medical school must teach medical ethics with respect for religious and other human values and their relationship to law and governance of medical practice. Students must be required to exhibit scrupulous ethical principles in caring for patients, and in relating to patients' families, others involved in patient care and to the community" (8). The standard regarding curriculum content (ED-20) is very specific about the requirement for students to demonstrate scrupulous ethical conduct and that instruction in such values must be undertaken before the student interacts with patients. "Scrupulous ethical principles imply characteristics like honesty, integrity, maintenance of confidentiality, and respect for patients, patients families, other students, other health professionals" (ED-20, 25). Importantly, as students progress through the curriculum, becoming increasingly more active in patient care, their adherence to ethical principles should be observed and evaluated as well as reinforced through formal instruction. |

(Appendix 6.1 table continues)

Appendix 6.1. Accreditation standards dealing with ethics: ACTT, BAC, CAAM-HP, CACET and UCJ (*continued*)

Agency	Standard
Caribbean Accreditation Council for Engineering and Technology	While not espousing core values or particular standards, CACET has outlined the following: *Declaration* (2008) Establish[ed] on the principles of advancing the quality of accredited programmes; demonstrating accountability; encouraging continuous self assessment and improvement; employing fair, equitable, transparent and publicly announced procedures; respecting variety and diversity in higher education; honouring the autonomy and academic freedom of programmes and their professoriate; and maintaining economical and minimally intrusive methodologies and practices. (CACET 2008, 1) *Accreditation Manual* (2009) 1.6 CACET is committed to fair, equitable and transparent processes and to mechanisms that lead to continued improvement in the operations of CACET and the programs it accredits. 14.1 CACET will consider all information concerning programs and institutions confidential, with the exception of the decision to accredit in any given calendar year. "Confidential information" includes all knowledge gained by CACET of the institution and program during the accreditation process, and the history of interactions between CACET and the institution/program.
The University Council of Jamaica	Standard 10 – Integrity. "The institution demonstrates adherence to the highest ethical standards in its dealings with students, faculty, staff, board of directors, accrediting agencies, and the general public. Through its policies and practices, the institution endeavours to exemplify the values it articulates in its mission and related statements."

Sources: The Accreditation Council of Trinidad and Tobago, Criteria for Institutional Accreditation; Barbados Accreditation Council (BAC), Handbook for Accreditation (2012); BAC Accreditation Standards (2010), Caribbean Accreditation Authority for Education in Medicine and other Health Professions (CAAM-HP), Standards for the Accreditation of Medical Schools in the Caribbean Community (2011); Caribbean Accreditation Council for Engineering and Technology, Accreditation Manual (2009) and Declaration (2008); The University Council of Jamaica: Standards for Institutional Accreditation (2010).

References

Ali, Azard. 2010. "Accept ACTT Degrees". *Trinidad and Tobago Newsday* (online), December 4. http://www.newsday.co.tt/news/print,0131924.html.
BAC (Barbados Accreditation Council). 2012. *Handbook for Accreditation*.
Beckles, Hilary McD. Anthony Perry and Peter Whiteley. 2002. *The Brain Train: Quality Higher Education and Caribbean Development*. Kingston: Board for Undergraduate Studies, UWI.
Bogue, E. Grady, and Robert L. Saunders. 1992. *The Evidence for Quality: Strengthening the Tests of Academic and Administrative Effectiveness*. San Francisco: Jossey-Bass.
CACET (Caribbean Accreditation Council for Engineering and Technology). 2008. Declaration. Adopted in San Juan, Puerto Rico, April 8.
———. 2009. Accreditation Manual. CACET.
Council on Education for Public Health. 2009. "Conflicts of Interest and the Accreditation Process". http://ceph.org/about/policies/conflict/.
Extavour, Mervyn, and Louis Whittington. 2011. "Independence of Quality Assurance Agencies vis-à-vis Different Stakeholders". Paper presented at INQAAHE 2011 Conference, Madrid, Spain. http://www.inqaahe.org/admin/files/assets/subsites/13/documenten/1303 463155_4-can-quality-assurance-agencies-in-developing-and-small-countries-support -and-defend-the-independence-of-the-institution-extavour-whittington.pdf.
Fain, Paul. 2013a. "Pima Community College's Deep Accreditation Crisis". *Inside Higher Education*, March 28. http://www.insidehighered.com/news/2013/03/28/pima-community -colleges-deep-accreditation-crisis.
———. 2013b. "Accreditation Crisis Hits City College of San Francisco". *Inside Higher Education*, July 6. http://www.insidehighered.com/print/news/2012/07/06/accreditation-crisis-hits -city-college.
———. 2013c. "Game Changer for CCSF". *Inside Higher Education*, August 14. http://www .insidehighered.com/news/2013/08/14/education-department-reprimands-accreditor -san-franciscos-community-college.
Government of Antigua and Barbuda. 2011. "Government Issues Statement on American University of Antigua (AUA) College of Medicine", March 4 (press release). Ab.gov.ag.
Hallak, J., and M. Poisson 2007. "Corrupt Schools, Corrupt Universities: What Can Be Done?" Paris, UNESCO, International Institute for Educational Planning.
Holder, Alex. 2011. "AUA Defends Accreditation after Media Attack". *Caribarena.com*, March 9. http://www.caribarena.com/antigua/education/96876-aua-defends-accreditation-after -media-attack.html.
Jaschik, Scott. 2013. "City of San Francisco Sues Accreditor". *Inside Higher Education*, August 23. http://www.insidehighered.com/quicktakes/2013/08/23/city-san-francisco -sues-accreditor.
Kelderman, E. 2013. "Department of Education Warns Accreditor That Sanctioned City College of San Francisco". *Chronicle of Higher Education*, August 14. http://chronicle.com/article /Department-of-Education-Warns/141109/?cid=at&utm_source=at&utm_medium=en.
Lederman, Doug. 2009. "Shot Across the Bow". *Inside Higher Education*, May 1. http://www .insidehighered.com/news/2009/05/01/antigua.

Margetson, Don. 1997. "Ethics in Assessing and Developing Academic Quality". *Assessment and Evaluation in Higher Education* 22 (2): 123–33.

National Assessment and Accreditation Council India and Commonwealth of Learning. 2006. *Quality Assurance in Higher Education: An Introduction*. Bangalore, India: National Assessment and Accreditation Council India and Commonwealth of Learning Assessment and Accreditation Council (June).

Nayebpour, Mohamad R., and Daryl Koehn. 2003. "The Ethics of Quality: Problems and Preconditions". *Journal of Business Ethics* 44: 37–48.

Perkins, Anna K., and Patrick S. Dallas. 2013. "Good Practices in Selecting Quality Assurance Evaluators: A Joint Analysis from Jamaica". *UWI Quality Education Forum* (January): 135–68.

Scott, Michael E. 2010. "Dimensions of Quality and Quality Assurance at Higher Education Institutions". *UWI Quality Education Forum* (January): 17–30.

Wilson, Nadine. 2013. "Video: Move to Register All Tertiary Institutions". *Jamaica Observer*, June 4. http://www.jamaicaobserver.com/news/Move-to-register-all-tertiary-institutions_14413346.

7 | Quality Assurance for Tertiary Level Technical and Vocational Education and Training for the Caribbean

HALDEN A. MORRIS

Quality assurance in its simplest form is providing stakeholders with guarantees that a product or service has met minimum established standards. Navaratnam and O'Connor (1993) agree with Oakland (1989) that "quality assurance is the prevention of quality problems through planned and systematic activities including the establishment of a good quality-management system and the assessment of its adequacy, the audit of the operation of the system, and the review of the system itself" (5). They also agree with Evans and Lindsay (1989) that "quality assurance refers to activities directed toward providing customers with products or services of appropriate quality" (16).

For several decades, quality assurance has become the norm in higher education because "it has become very important for its stakeholders" (Mishra 2006, 24) and because of the vast expansion in the number of providers and the various levels at which they operate internationally. With the growing momentum and interest in technical and vocational education and training (TVET) at the tertiary level, quality is currently regarded as an essential component in the provision at this level and type of education. The need to provide quality education is a result of increased level of public scrutiny, greater demand from clients/students, requests from other institutions to facilitate transfers and matriculations, and the increased global perception of the role of TVET in sustainable national development (Navaratnam and O'Connor 1993). In recent times, funding organizations are requiring more detailed justification for the expenditure incurred, and students are demanding better value for their money. Additionally, tertiary level institutions are requiring quality assurance information from each other prior to employing staff from such institutions or facilitating student transfers from one institution to another. Mishra (2006) identified competition, consumer satisfaction, maintaining standards, accountability, improved employee morale and motivation, credibility, and image and visibility as reasons everyone should be conscious about quality education. This chapter clarifies the minutely researched field of quality assurance for tertiary TVET and also provides information on a successful quality assurance approach being employed internationally. It also presents indicators of quality for TVET at tertiary levels and analyses these indicators in relation to what exists in the Caribbean TVET context.

Tertiary Level TVET

The 2001, United Nations Educational, Scientific and Cultural Organisation and International Labour Organisation general conference on TVET referred to the discipline as covering "those aspects of the educational process involving, in addition to general education, the study of technologies and related sciences, and the acquisition of practical skills, attitudes, understanding and knowledge related to occupations in various sectors of economic and social life" (UNESCO and ILO 2002, 7). Tertiary level TVET includes curricula pursued in technical institutes, colleges and universities such as the University of the West Indies; the University of Technology, Jamaica; the Mico University College; and the University of Trinidad and Tobago, to name a few Caribbean institutions. These institutions facilitate teaching–learning in order to attain advanced qualifications and professional certification in technical and vocational areas. However, according to Mishra (2006), not all TVET curricula pursued in these institutions can be classified as "tertiary level TVET". At the tertiary level, students should be exposed to advanced knowledge and understanding as well as skills needed to address critical developmental issues, usually leading to at least an associate degree or its equivalent.

Mishra (2006) shares the view that higher education is about imparting in-depth knowledge and understanding so as to advance students to new frontiers of knowledge in different walks of life. He further claims that higher education "develops the student's ability to question and seek truth and makes him/her competent to critique on contemporary issues. It broadens the intellectual powers of the individual within a narrow specialization, but also gives him or her a wider perspective of the world around" (5). To emphasize this thought, Mishra introduced Barnett's (1992) four prominent concepts of higher education as: (1) the production of qualified human resources, (2) training for a research career, (3) efficient management of teaching provision and (4) a matter of extending life chances. These concepts are indeed congruent with those of tertiary TVET and further endorse the role that tertiary level TVET plays in society.

Tertiary level TVET provides the basis for scientific and technological advancement and economic growth of a country. This education is responsible for development and certification of the workforce in all sectors of the society, namely tourism, agriculture, mining, manufacture, transportation, construction, and information and communication technologies, among others. Tertiary level TVET provides opportunities for lifelong learning which allows persons to upgrade their knowledge or skills, ensuring currency in industry practice. It also facilitates access to training in other linked areas as needed by the individual or the society, upgrades leadership skills to provide quality leadership in areas of specialization, and promotes quality and social justice by focusing on the values and attitudes needed for developing individuals and the society (Navaratnam, and O'Connor 1993).

Quality Assurance for Tertiary Level TVET

TVET at the tertiary level is not static; therefore, quality assurance mechanisms must assure ongoing improvement in the institutions' operations as they seek to produce quality outcomes while strategizing to satisfy emerging trends in the society. During the process of improvement, quality needs be at the forefront. All stakeholders must make quality a key priority, know and perform their roles and responsibilities, and adhere to established quality assurance principles.

Quality assurance in TVET can be established based on internal standards, external standards, or a combination of both. However, it is critical to establish and operate a quality assurance system that complies with recognized external standards, such as the ISO 9001 series or the Accreditation Board for Engineering Technology (ABET). Such actions are unequivocal proof of the institutions' commitment to quality tertiary level TVET and provide objective evidence of their robust internal quality assurance systems.

The establishment of an effective quality assurance process for tertiary level TVET can be a major challenge, since this requires commitment from all stakeholders to the process. This process involves the introduction of quality concepts that requires a new outlook and thinking in the management and delivery of the courses and programmes. The institution will have to make considerable commitments with respect to the provision of management and resources that are required to attain and maintain the standards that are acceptable to both industry and the educational institution. Additionally, it may be necessary to develop a customized quality assurance process that is relevant and feasible for the specific nature of the TVET programme and the conditions found in the institution.

The commitment of the management of TVET institutions is crucial to the success of any quality assurance initiative at the tertiary level. All managers (both senior and line) must know what the stakeholders want and focus on the goal of delivering programmes to meet those expectations. All groups and individuals must be attuned to the pulse of the institution and must have a clear understanding about the organization. To accomplish this, Navaratnam and O'Connor (1993) suggested that the institution should develop a vision and a mission for quality and establish management commitment to standards and for quality performance. Strategies for implementation should then be chosen and the staff educated about the process. The process and progress should be audited and plans put in place to make continuous improvement. They emphasized the fact that quality assurance is a team activity and that it must have the support and contribution of everyone in the organization.

The body responsible for quality assurance of TVET in Jamaica, the National Council on Technical Vocational Education and Training (NCTVET), embraces the strategies detailed by Navaratnam and O'Connor (1993) to accomplish quality targets. These strategies are, however, deployed primarily at the lower levels of TVET since the NCTVET concentrate at these levels. In the Caribbean, tertiary level programmes are evaluated by the University Council of Jamaica, the Barbados Accreditation Council, the Accreditation Council of Trinidad and Tobago, National Accreditation Board of the Commonwealth of Dominica, the Antigua and Barbuda National Accreditation Board, and the National Accreditation Council of Guyana. These bodies are members of a Caribbean network of accreditation bodies, the Caribbean Area Network for Quality Assurance in Education, which was established in 2003 "to promote best practices in quality assurance and quality enhancement in the region through the capacity building, sharing of information about the maintenance, evaluation, accreditation and improvement of tertiary/higher education" (www.canqate.org).

Quality Assurance Models and Approaches

In TVET, there are basically two quality assurance models – the compliance model and the evaluative model. In the compliance model, minimum outcomes, targets and standards are established and entities are expected to meet these benchmarks. This is intended to ensure national consistency in the provision and assessment of TVET. This model is expensive to implement and requires strong centralized systems with regular reviews and audits. Sanctions are usually employed in the compliance model when organizations fail to meet the required standards. In the evaluative model, optimum standards are established, against which performances are judged. This differs from the compliance model as it employs a systematic process of enquiry, designed to provide independent judgements about the organization's performance and capability to deliver quality education.

The evaluative model is considered superior and it is usually adopted after a period of operating under the compliance model, where organizations have become accustomed to being scrutinized. Bowen-Clewley, Cooper and Grannall (n.d.) suggested that this model is suited for countries in which qualification levels and relations, such as a national qualifications framework, have been established and the countries have had experience with the compliance model. For countries to effectively implement this model, a strong commitment to quality assurance and continuous improvement is important. Public understanding of the quality assurance processes and the need for such a mechanism is also necessary.

The Quality Assurance Process

Quality assurance can be approached from several perspectives; however, many educational institutions have adopted an instrumental approach in which quality is seen as something that fits its purpose or mission in accordance with publicly accepted standards of accountability and integrity. In this regard, the quality assurance system is effective if it achieves this purpose. The basic instrumental approach requires the educational institution to establish its mission and introduce strategies and management processes in order to achieve the mission. Additionally, the institution is required to use qualitative and quantitative performance indicators to show its progress and monitor its ability to carry out its established functions. Finally, the institution is required to implement what it has learned in its action plans (Lim 2009).

Three methods employed in quality assurance assessment in TVET include an a priori method in which the institution employs established management and academic committees with external members. In this method, an external examiner is employed to provide external benchmarks and to carry out the quality assurance functions. However, this method is often characterized as inefficient since the outcomes of assessment are inconsistent as they rely heavily on the background of the external examiner. Additionally, the method is viewed as inadequate because of the lack of empirical evidence to support the claims. Second is the *stepwise backtracking assessment* method, which is basically a combination of the a priori and empirical evidence. Performance indicators are used to measure the effectiveness of the management process for this method, thus making it sound and logical. Notwithstanding, there are problems with using the performance indicators since these are not always reliable. In some instances, indicators are based on situations that are dissimilar to that being evaluated. Additionally, outcomes can be tailored to meet the expectations of the indicators. The third method is the *external evaluation method* in which quality assurance is provided by an independent quality assurance organization such as the University Council of Jamaica or NCTVET in Jamaica, Barbados Accreditation Council in Barbados, or Accreditation Council of Trinidad and Tobago.

Lim (2009) explained, by quoting the National Committee of Inquiry into Higher Education (1997) and Alderman (1996), that the usefulness of the external evaluation method is questionable "since institutions have learned to play the system effectively and succeeded in obtaining higher rating than they deserve by employing consultants to provide advice on tactics and strategies" (185). He further explained that in some instances, institutions send their key players to attend professionally executed short courses where they learn the unwritten rules of the game and strategies on how to present aims and objectives in quality portfolios, and provide evidence to support arguments. Despite the shortcomings identified, all the methods are employed in the evaluation process based on the perceived benefits of each method.

Indicators of Quality for Tertiary Level TVET

Indicators of quality, sometimes referred to as performance indicators, are qualitative and quantitative measures used in quality assurance to indicate how the organization is performing against various measures. There have been several criticisms of the use of performance indicators. A common one is that the indicators are usually out of date since they are based on ex post data, as they refer to what has been done and not on what should be done in the future. Some criticize that quality indicators evaluate the targets set to be achieved without evaluating the process used to achieve the targets because they focus on outcomes. Relevant indicators depend on the coverage and scope of public and private funding of the various types of TVET programmes.

Vocational Training Council of Hong Kong

An example of performance indicators is taken from the Vocational Training Council of Hong Kong, which adopted a quality assurance system in order to remain competitive and to satisfy requirements of the Hong Kong Institution of Engineers. The Hong Kong quality assurance system has four parts and functions, namely, "(i) quality policy, which determines the institution's commitment to provide quality; (ii) quality assurance framework, which sets out the framework for implementing enabling management processes and the performance indicators to measure their effectiveness; (iii) an evaluation system, which measures the impact of the enabling management processes on the provision of quality learning and teaching, and (iv) an internal monitoring system, which tracks the institution's ability to implement improvement plans" (Lim 2009, 187). The quality assurance framework was introduced in 2000, modelled on the Malcolm Baldrige Educational Criteria for Performance Excellence. This framework was later adjusted to incorporate elements

Table 7.1 Quality assurance framework for Hong Kong

Component	Subcomponent	Performance indicator
Driver	Leadership	Leadership at various levels of the operational unit
Enabler	Strategic planning	Strategy development and implementation
	Financial resources	Financial planning and budgetary control
	Human resources	Human resources management and development
	Educational and support process	Programme design and development
		Teaching and learning
		Educational support
Result	Student performance	Retention/completion rates
		Pass rate
	Stakeholder satisfaction	Student/trainee/internal client satisfaction
		Staff satisfaction
		Employer/industry satisfaction
	Budgeting and finances	Unit cost
	Organizational effectiveness	Number of planned TVET places
		Enrolment rate
		Employment rate

of the European Foundation Excellence Model and the Singapore Quality Award Framework, which resulted in the adoption of a Plan–Do–Check–Act quality cycle; this latter framework, to a great extent, conforms to the ISO 9001 quality management system. The framework consisted of three components, arranged in three areas, referred to as the driver, the enabler and the result, as outlined in table 7.1.

The driver provides leadership and interfaces at all levels of the system. The enabler is responsible for making things happen through strategic planning, provision of financial and human resources and other educational resources. Results address student performance, stakeholder satisfaction, organizational effectiveness and other outcomes measures.

Inter-Agency Group on TVET

In 2009, the United Nations Educational, Scientific and Cultural Organisation, the Organisation for Economic Co-operation and Development, the World Bank, the International Labour Organisation, the European Commission, the European Training Foundation and the Asian Development Bank came together and established

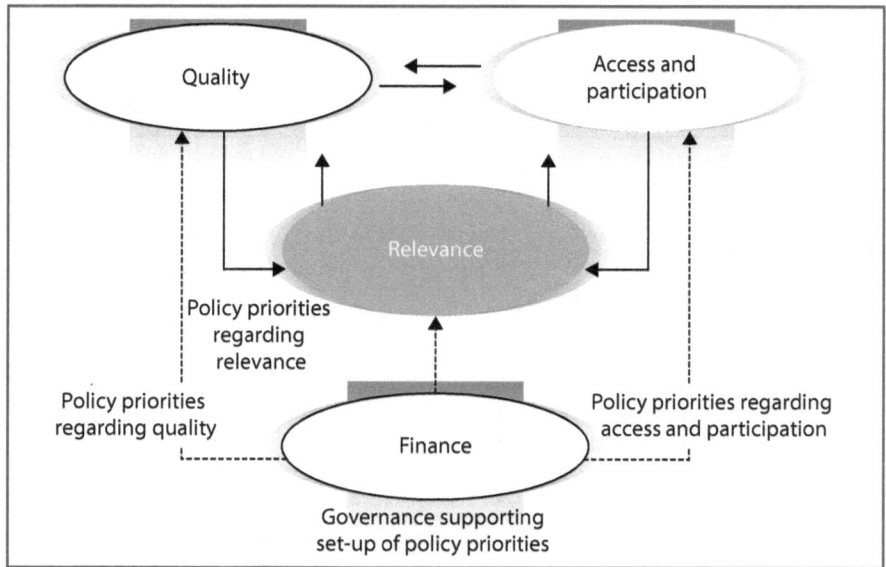

Figure 7.1. Conceptual framework for selected policy areas (IAG-TVET) (UNESCO 2012)

a working group called the inter-agency group on TVET (IAG-TVET) (European Training Foundation (ETF), United Nations Educational, Scientific and Cultural Organisation (UNESCO), and International Labour Organisation (ILO)) (UNESCO 2012). This group was tasked with the responsibility of developing indicators of quality TVET that can support developing countries through assessing the efficiency and effectiveness of their national TVET systems. In order to accomplish this, the group established a conceptual framework (see figure 7.1) that was based on the assumption that policymakers and stakeholders have to optimally combine priorities related to components of relevance, equity and quality, within available financial resources, in the context of institutional settings and governance. These indicators, though not intended for specifically tertiary TVET, set the stage for further work towards the development of indicators of quality for the higher educational levels in TVET.

Table 7.2 provides a synopsis of the resulting indicators of quality, which are concentrated within the components defined in the inter-agency group on TVET framework.

Accreditation Board for Engineering Technology

Another example of a quality assurance approach is the system and processes employed by the ABET, a primary organization responsible for monitoring, evaluating and

Table 7.2. Categories and types of indicators used in TVET evaluation

Area	Indicators and data availability		
	Data readily available	Data not readily available	Data often not available
1. Financing	1.1 Spending in formal TVET	1.2 Total TVET spending by student	1.3 Share of companies providing apprenticeship and other types of training (by size of the company) 1.4 Share of apprenticeship and other types of training spending in labour cost (by size of the company)
2. Access *Access as opportunities* *Access as participation* *Access as transition*	2.1 Enrolment in vocational education as a percentage of total enrolment in the formal education system 2.2 Enrolment by type of TVET programme	2.6 Typology of admission policies to formal school-based TVET 2.7 Transition paths from upper secondary TVET education	2.3 Work-based learning participation rate 2.4 Equity 2.5 Unsatisfied demand for TVET 2.8 Policies on articulation with schooling/higher education
3. Quality and innovation	3.1 Student/teacher ratio in formal TVET and in general programmes 3.2 Completion rate in TVET programmes and in general programmes	3.3 Share of apprentices completing registered programmes as a percentage of all apprentices starting registered programmes 3.4 Share of qualified teachers in TVET and in general programmes	3.5 Relevance of quality assurance systems for TVET providers 3.6 Investment in training of teachers and trainers 3.7 Utilization of acquired skills at the workplace 3.8 Share of ICT training activities in TVET 3.9 Satisfaction of employers with TVET graduates
4. Relevance	4.1 Employment to population ratio 4.2 Unemployment rate 4.3 Employment status 4.4 Employment by economic sector 4.5 Employment by occupation 4.6 Literacy rate	4.7 Informal employment rate 4.8 Time-related unemployment rate	4.9 Working poverty rate 4.10 Average real earnings 4.11 Number of vacant jobs 4.12 Net job creation 4.13 Youth outside labour force 4.14 Discouraged workers

Source: ETF (2012).

certifying the quality of education in applied science, computing, engineering and technology in the United States. According to ABET (2010), the body organizes and carries out a comprehensive process of accreditation of pertinent programmes leading to degrees in these areas. The board also assists institutions in devising stimulating innovation and planning their educational programmes. Programmes are evaluated against a set of established criteria which are based on the principles of continuous quality improvement. The criteria are intended to ensure the quality of educational programmes, foster the systematic pursuit of quality improvement of these educational programmes and to help educational programmes that satisfy the needs of constituencies in a dynamic and competitive environment. It is the responsibility of the institution seeking accreditation to demonstrate clearly that the programme meets the criteria.

The first criterion is general and addresses students, programme educational objectives, programme outcomes, continuous improvement, curriculum, faculty, facilities and support. The second is programme specific and covers areas that are specific to the programme being evaluated. ABET has gained recognition internationally and has been called upon by several countries, including those in the Caribbean, to provide quality assurance services. In fact, it has become the default accreditation agency for many countries. In addition to full compliance, ABET employs four categories of compliance in their evaluation. These are defined in table 7.3.

It is evident from the foregoing examples that indicators of quality for tertiary level TVET are diversified across quality assurance agencies; however, they incorporate essential elements to assure quality across a wide cross section of TVET

Table 7.3. ABET's categories of compliance

Categories of compliance	Description
Concern	Indicates that a programme currently satisfies a criterion, policy or procedure. However, the potential exists for the situation to change such that the criterion, policy or procedure may not be satisfied.
Weakness	Indicates that a programme lacks the strength of compliance with a criterion, policy or procedure to ensure that the quality of the programme will not be compromised. Remedial action is required to strengthen compliance with the criterion, policy or procedure prior to the next evaluation.
Deficiency	Indicates a criterion, policy or procedure is not satisfied. The programme is not in compliance with the criterion, policy or procedure.
Observation	This is a comment or suggestion that does not relate directly to the accreditation action, but is offered to assist the institution in its continuing efforts to improve its programmes.

specializations. A common thread that is evident among the quality assurance systems examined is a mechanism for risk mitigation and management, which can be further explored for individual systems. The critical factors that determine the outcomes rest with the seriousness of the personnel who are charged with the responsibility of providing evidence to support the claims of meeting quality targets and the knowledge, skills and understanding of personnel employed to carry out the evaluation processes. Invariably, for the latter, persons are trained by the agency prior to their involvement in the evaluation process.

Quality Indicators for Tertiary TVET in the Caribbean

Quality assurance for tertiary level TVET in the Caribbean context is still in its infancy. Indeed, there are no specifically established indicators for quality for tertiary TVET, except for the general indicators employed by University Council of Jamaica, Barbados Accreditation Council, Accreditation Council of Trinidad and Tobago and other Caribbean quality assurance agencies; more importantly, these indicators were not necessarily established for tertiary level TVET. NCTVET, for example, has developed indicators of quality for TVET programmes in Jamaica; it has made these indicators available for adaption by other Caribbean quality assurance bodies. These indicators, however, do not specifically target the national TVET system but focus on areas of quality assurance and control, staff resources, physical resources, client services and activities, training, assessment, and evaluation for all aspects of a general education system that delivers technical and vocational training.

Indicators of quality are influenced by the model employed in the quality assurance process. To this end, the evaluative model is recommended for tertiary level TVET in the Caribbean since there is familiarity with quality assurance at the secondary level of the education system. Quality indicators for tertiary level TVET in the Caribbean would therefore include aspects of those established internationally such as those used by the Vocational Training Council of Hong Kong or those used by ABET. Those established by the NCTVET and other quality assurance agencies in the Caribbean would be of great value as well. The quality indicators incorporated in the University Council of Jamaica's model, which includes the quality of the students, the quality of the faculty, the programme, assessment methods, resources and student support services, are appropriate. The University Council of Jamaica claims that one of its primary functions is to assure educational quality and institutional integrity to the institutions, the students and the public. The Barbados Accreditation Council, Accreditation Council of Trinidad and Tobago, other Caribbean quality assurance agencies and Caribbean Area Network for Quality Assurance in Education have established several standards and outlined detailed indicators of quality for these standards; however, these are not specifically configured for TVET.

There is no doubt that the foregoing general indicators are appropriate for tertiary level TVET; however, it is imperative that consideration be given to unique aspects of TVET that are not normally considered in general education. Criteria that focus on the imperatives of TVET such as management of industrial processes, meeting industrial and commercial standards, and industry involvement in the curriculum development and delivery processes are essential. Inclusion of new and modern technologies such as information and communication technologies, simulators, global positioning systems and the utilization of new manufacturing processes will impact quality assurance. Additionally, with the introduction of new skills and competencies such as those required by logistics hubs, ship maintenance/ dry dock facilities, hydroponics and other advanced technical areas, an increase in demand for training in higher order skills and productivity will require a new paradigm for assuring quality tertiary TVET in the Caribbean.

Conclusion

Quality assurance for tertiary level TVET has become essential in recent years because of the emergence of global competitiveness and other factors such as changing stakeholder demand. Many developing countries are increasingly realizing the potential role that this aspect of education can play in their national development; as such, these countries are creating national TVET strategies that take the current societal quality assurance trends into account. Quality assurance for TVET has been given significant attention in the Caribbean since all quality assurance bodies have included this aspect of education as part of their remit. The University Council of Jamaica, Barbados Accreditation Council, Accreditation Council of Trinidad and Tobago as well as all other quality assurance agencies in the Caribbean have employed the external evaluation method since their inception to assure quality in various areas of education inclusive of tertiary level TVET. The NCTVET has also employed this method quite successfully for various programmes offered in TVET training institutions and in some secondary schools. It has been recognized that high-quality TVET is essential in positioning Jamaica and the rest of the Caribbean to contribute in a significant manner to global development, participation and interaction.

High-quality TVET facilitates appropriate human resource development within the context of the particular culture and thus higher career mobility through focusing on competence development, as opposed to just knowledge acquisition. In a quality assured system, students are provided an opportunity to optimize their talents and capabilities, to develop knowledge, technical skills, and leadership skills that are compatible with the needs and aspirations of the society and to participate effectively in the modern knowledge economy. However, given that tertiary TVET has not been given significant attention, developing and implementing an effective quality assurance mechanism for this level of TVET in the Caribbean is imperative. Doing so will provide opportunities for TVET practitioners to gain access to and benefit from the international TVET community while ensuring that the quality of the training and education be maintained and continually improved.

References

ABET (Accreditation Board for Engineering and Technology). 2010. "Program Evaluator Training Manual". http://www.abet.org.

Barnett, R. 1992. *Improving Higher Education: Total Quality Care*. Buckingham: SRHE & Oxford University Press.

Bowen-Clewley, L., Karen Cooper and Ray Grannall. N.d. "A Comparison of Quality Management Approaches for the Training and Vocational Sector in Seven Countries". http://www.iaea.info/documents/paper_4d2287df.pdf.

European Training Foundation (ETF), United Nations Educational, Scientific and Cultural Organisation (UNESCO), and International Labour Organisation (ILO). 2012. "Proposed Indicators for Assessing Technical and Vocational Education and Training Inter-Agency Working Group on TVET Indicators". UNESCO. April. http://www.etf.europa.eu/webatt.nsf/0/E112211E42995263C12579EA002EF821/$file/Report%20on%20indicators%20April%202012.pdf.

Evans, J.R., and W.M. Lindsay. 1989. *The Management and Control of Quality*. St Paul, MN: West Publishing.

Lim, David. 2009. "Testing the Effectiveness of a Quality Assurance System: The Example of Hong Kong". *Journal of Vocational Education and Training* 61 (2): 182–202.

Mishra, Sanjaya. 2006. *Quality Assurance in Higher Education: An Introduction*. Bangalore, India: National Assessment and Accreditation Council (NAAC); Vancouver: Commonwealth of Learning.

Navaratnam, K.K., and Rory O'Connor. 1993. "Quality Assurance in Vocational Education: Meeting the Needs of the Nineties". *Vocational Aspects of Education* 45 (2): 113–32.

Oakland, J.S. 1989. *Total Quality Management*. New York: Nicholas.

UNESCO. 2012. "Technical and Vocational Education, Part 1: Conclusions of the Third International Congress on TVET". Executive Board 190th Session. Paris. http://unesdoc.unesco.org/images/0021/002175/217544e.pdf.

UNESCO and ILO. 2002. *Technical and Vocational Education and Training for the Twenty-first Century: UNESCO and ILO Recommendations*. Paris: UNESCO-UNEVOC Publications. http://unesdoc.unesco.org/images/0012/001260/126050e.pdf.

8 | A Flexible Model for Quality Assurance in Technical and Vocational Education and Training in the Caribbean

PAULETTE J. DUNN-PIERRE

Having spent many years in the doldrums, technical and vocational education and training (TVET) is now regaining much attention in developed and developing countries because of the invaluable contribution it has been found to make in improving the competitiveness of the workforce (ILO 2010). The major change in the overhaul of TVET has occurred in the area of quality assurance, which in the past did not always guarantee consistency in TVET training and delivery as competency standards were little used in developing learning outcomes (UNESCO-UNEVOC 2013). Quality assurance in TVET refers to the monitoring of training, assessment and certification of individuals in keeping with stipulated requirements. The quality assurance system helps TVET institutions gain confidence in outputs and assists in the promotion of and increase in public trust. Quality assurance requires established benchmarks against which qualifications, courses and providers can be assessed. In the Caribbean, quality assurance in TVET falls under the auspices of the Caribbean Association of National Training Agencies (CANTA), a group comprised of national TVET councils, national training agencies (NTAs) or TVET focal points within ministries of education.

This chapter provides an overview of the CANTA standards for quality assuring the TVET system in the region to award the Caribbean Vocational Qualification (CVQ); it also identifies the reasons behind the slow uptake of countries in developing and implementing the CANTA standards. A case is made for a review of the current approach to the strategies used by NTAs, TVET councils and other TVET focal points charged with the responsibility for planning and coordinating TVET at the local level. It concludes by making recommendations for a flexible quality assurance compliance model that countries can introduce incrementally given their resource capabilities.

Although CANTA is a voluntary association, it is recognized by the Caribbean Community Secretariat and is charged with facilitating "the uniform provision of a competency-based training, assessment and certification system throughout the region" (CANTA 2006, 7). The association, with the aid of two committees (Quality Assurance and Standards), conducts workshops and seminars throughout the region (funded currently by the Canadian International Development Agency) and assists its members in standardizing and implementing a standards-driven, competency-based approach to workforce development and certification, thereby supporting an improved Caribbean workforce and facilitating the movement of skilled workers across the region, one of the main pillars of the Caribbean Single Market and Economy.

The definition of key terms related to quality in TVET institutions varies among countries and regions. For the purpose of this discussion, the following definitions are used in the context of quality assurance in the Caribbean Community region:

Technical and vocational education and training (TVET) is broadly defined as education that facilitates the acquisition of practical skills, know-how and understanding, necessary for employment in a particular occupation, trade or group of occupations. The United Nations Educational, Science and Cultural Organisation defines TVET as "those aspects of the educational process involving, in addition to general education, the study of technologies and related sciences, and the acquisition of practical skills, attitudes, understanding and knowledge relating to occupations in various sectors of economic and social life" (UNESCO 2010, 5). As such, TVET involves a wide range of learning experiences that enable people to become more productive in the world of work.

Quality in TVET refers to maintaining a particular standard of best practice. For the Caribbean, these standards are developed by quality assurance experts, educators and industry practitioners, and they govern three general areas: (1) TVET qualifications, that is, national vocational qualification and the CVQ; (2) TVET courses based on occupational and or competency standards and (3) training institutions or training providers. An improvement in quality results in a better perception of TVET and its contribution to economic development.

The *regional vocational qualifications framework* is a quality-assured system for the development, recognition and award of qualifications based on competency standards of knowledge, skills and attitudes acquired by learners or workers adhered to by TVET institutions across the Caribbean Community region. The framework (appendix 8.1) links together all the qualifications across the region and promotes lifelong learning within a seamless education and training system. It is composed of titles, five occupational levels and guidelines that define each qualification, together with principles and protocols covering articulation and issuance of qualifications and statements of attainment or statements of competency. The key objective of the regional vocational qualifications framework is to provide the basis for housing mutually recognized national and regional qualifications.

Quality Assurance in TVET and Development

TVET is acknowledged as a means of transforming economies and empowering individuals and communities through the development of skills and is recognized in the Caribbean Community region as a priority area for development intervention as is reflected in the recently revised regional TVET strategy document (CANTA 2012). Developments in the last three decades have made the role of TVET more decisive: the globalization process, technological change and increased competition due to trade liberalization necessitating higher skills and greater productivity among workers in all sectors of the economy. Skills development, which takes place in technical and vocational training, encompasses a broad range of core skills (entrepreneurial, communication, financial and leadership), and enables individuals to be equipped for productive activities and employment opportunities. Quality assurance therefore becomes an important element in TVET as it lends credibility and engenders trust by the public in the standard and quality of training in TVET training institutions.

TVET has therefore become an integral aspect of the reform of education in many developing countries, including the Caribbean. One of the most important features of TVET is its orientation towards the world of work and hence there is an emphasis in the TVET curriculum on the acquisition of employable skills to develop competent workers at different levels within the regional vocational qualifications framework.

However, developing countries in particular are struggling to keep up with the pace of drastic changes as they lack the requisite financial and human resources necessary to remain current and to adjust to the constantly changing global situations. This results in a widening of the skills gap between Caribbean workers and their global counterparts. Conversely, in developed countries, there has generally been a reduction in the demand for unskilled labour and a rise in the market value of advanced skills and workplace competencies.

The ultimate challenge lies in keeping abreast with technological change but globalization continues to have a major influence on the need for flexible work skills. More of the same secondary education and skills training, of indifferent quality, is no longer acceptable.

Arguably, the many benefits claimed for TVET, such as higher productivity, readiness for technological change, openness to new forms of work organization and the capacity to attract foreign direct investment, all depend on the quality of the skills acquired and a dynamic environment in which they can be applied. Developmental funding agencies such as the Caribbean Development Bank, the World Bank and the Canadian International Development Agency are now calling for improved quality in technical and vocational training, and NTAs in the Caribbean are responding to this call through CANTA.

Virtually all Caribbean countries have implemented, or are in the process of implementing, reforms in their education and training systems (Education Transformation Programme [Jamaica]; Pillars for Partnership and Progress: The OECS Education Reform Strategy 2010) in an effort to create societies that foster social and economic development. Caribbean Community member states are developing and establishing quality assurance mechanisms guided by CANTA standards in an effort to assist post-secondary TVET institutions become teaching and learning organizations in which quality improvement is continuous. The use of approved standards for the delivery of curricula include

- the use of regional assessment standards
- competence of teachers and trainers
- competence of TVET institutional managers
- personal appraisal systems and identification of training needs of teachers and trainers
- internal quality assurance and self-assessment undertaken by TVET providers
- guidance and support services for learners/trainees
- learning and teaching equipment and resources
- reliable and timely data about TVET activity and outputs
- collaboration and partnerships with employers and other stakeholders
- links with parents and others who might influence the decisions of potential learners (http://www.unevoc.unesco.org)

CANTA Quality Standards

The regional approach to training, assessment and certification became a reality with the formation of CANTA in 2003. The Human, Employment and Resource Training Agency (HEART-NTA), Jamaica; the TVET Council Barbados; and the National Training Agency of Trinidad and Tobago were founding members. Today these three entities play a crucial role in facilitating knowledge-sharing and promoting quality assurance within member states, particularly those preparing for full implementation of the CVQ. The CVQs are work-based qualifications derived from internationally bench-marked occupational standards against which the skills of the individuals are assessed. CANTA plays a major role in the region by ensuring that the requirements for training, assessment and certification are met by enterprises and other training providers. This section briefly describes CANTA quality standards used in national TVET systems.

The quality assurance strategies as delineated by CANTA "provide a framework for the delivery of a credible qualification, which certifies the attainment of world-class standards" (CANTA 2009, 6). An assurance is given that the strategies will be "implemented consistently and will be added to, reviewed and updated in order to ensure that the desired quality product is delivered, a world-class workforce" (CANTA 2009, 6). Each member state is charged with developing the mechanism to award the CVQ by adherence to the quality assurance standards, which are applied in quality audits. The five areas where the principles of quality assurance are applied in the process of training, assessment and certification are:

1. development of training outcomes that reflect workplace requirements (occupational standards)
2. training delivery
3. assessment
4. certification
5. adherence to best practice by the awarding body

Standards

Occupational standards are statements about the knowledge, skills and attributes that individuals need to perform in the workplace. Developed by industry practitioners, they form the basis for ensuring that training and assessment meet the needs of industry and learners are assessed against them for the award of the national vocational qualifications and Caribbean Vocational Qualifications.

Training Delivery/Approval of Assessment Centres

The second major area of quality assurance is in the delivery of training. This can be work- or institutional-based, distance or a mix of training modalities. In each case, standards are defined for the delivery of training and assessment, and providers are required to be approved by the awarding body. Approval criteria cover:

- management of information (record management, security and storage)
- staff resources
- physical resources

- learning resources
- statutory compliance
- internal verification procedures
- assessors
- awarding bodies are responsible for approving centres to deliver CVQs

Assessment

The third area of quality addresses assessment within the certification framework (appendix 8.2). Assessment is defined as "the process of collecting evidence and making judgements about whether or not competence has been achieved when measured against the occupational standards" (CANTA 2009, 4). Assessment instruments are validated and assessors are trained and certified in competence-based assessment methodology. In order to be registered as a CVQ assessor and be allowed to assess competence, individuals must be trained and certified at CVQ Level 4 training and assessment. Certification enables assessors to consistently assess the performance of learners, thereby enhancing the quality outcomes achieved by TVET graduates. CVQ awarding bodies and assessment centres are also required to adhere to internal and external verification guidelines.

Certification

Fourth, candidates are issued a CVQ from the local awarding body, where applicable, on the basis of having met the requirements for certification. This is done on the basis that the training/assessment centre (1) subscribes to the regional vocational qualifications framework (appendix 8.1), (2) has adhered to the standards-driven, competence-based approach to training and assessment and (3) is subject to the quality assurance principles under the auspices of the local NTA, focal point or accreditation body.

Awarding Body Best Practice

Finally, the quality assurance process for the CVQ is implemented and monitored by the NTA or the TVET councils, and CANTA requires that the awarding body demonstrates international best practices in training and delivery in the areas of leadership and management; awarding and assessment of candidates; customer service; design and development of qualifications; promoting diversity and inclusion; and continuous improvement.

Observations

Even though quality standards have been developed by CANTA, member states have experienced challenges accessing the standards as the website has not kept up with the pace of the demand and therefore have been slow in implementing them. The guidelines that are required by NTAs (for example, the *Assessment Guidelines for the CVQ* and *External Verification Guidelines for the CVQ*) are not easily accessible. In one study, it was observed by TVET practitioners that not enough guidance was provided by CANTA and there was some uncertainty as to what is to be done (Dunn-Smith 2012). In fact, only two of the member states – St Lucia and Grenada – have applied and have been audited by CANTA in the ten-year history of the association. This was in 2012, as prior to then, their quality assurance systems and structures had not been sufficiently developed owing to the lack of human, physical and technical resources.

An assumption has been made (rightly or wrongly) that the older NTAs, namely, HEART-NTA (Jamaica), the TVET Council (Barbados) and the NTA (Trinidad and Tobago), have met and are adhering to all of the standards and requirements that these agencies have developed, assessed and approved. For transparency and credibility and to dispel any doubts as to the management of quality in the member states with regard to adherence to standards, all NTAs and TVET councils awarding CVQs should themselves be subject to regular audits monitored by members drawn from CANTA as well as from similar external awarding bodies in the region as is practised in other jurisdictions such as Australia and the United Kingdom.

A Developmental Approach

Given the length of time it has taken for the other member states to implement quality assurance strategies and to award the CVQ, it seems fair to conclude that this situation has arisen owing to inadequate human, physical, technological and financial resources. To this end, a developmental approach is recommended for small and emerging NTAs to develop and implement their quality assurance mechanisms for the training and subsequent certification of their workforce. Technical assistance (using outsourced short-term consultancies) should be provided by CANTA and networks strengthened through greater use of technology. To assist the emerging (and mature) NTAs in developing their mechanism to award the CVQ, the quality assurance strategy recommended by UNESCO-UNEVOC should be adopted. The UNESCO-UNEVOC strategy consists of four main components:

1. a compliance model
2. a method for assessment and review

3. a monitoring system
4. a measurement tool

This model addresses in some fashion the question of cost, accessibility of standards and human resources.

The Compliance Model

The compliance model is an input system that is intended to ensure national consistency in the provision and assessment of TVET and focuses on five major activities:

1. Establishing standards and criteria to be applied for registration of standards, qualifications, training providers, assessors and/or courses;
2. Developing processes for ensuring consistency of assessment both within and between providers, assessors and/or courses;
3. Developing an internal audit requirement within providers;
4. Placing a strong emphasis on independent external audit to identify areas of compliance and non-compliance; and
5. Implementing processes to ensure remediation of non-compliance

This compliance model can be followed step-by-step over time, allowing smaller, less-resourced NTAs to strengthen their capacity, competence and capability as they go along. It also allows for countries to identify where their weaknesses and strengths lie, so as to be better able to make the requisite interventions with the assistance of CANTA. The disadvantage is that the model is high in cost as it requires strong centralized systems and regular reviews and audits, with follow-up of non-compliant performance. This is where CANTA should be able to intervene with the assistance of the more mature NTAs in the association. Although costly, this approach is necessary as TVET has not had a tradition of quality assurance in the region. This model is most suitable in member states where (1) there are low or uneven levels of quality provision of TVET, (2) there is a lack of consistency between courses and problems with parity of esteem of those courses and (3) a variety of training providers operate within agencies with different organizational structures and requirements (Bowen-Clewley, Cooper and Grannall 2010).

Evaluation/Assessment and Review

A method of assessment for evaluation and review has been stipulated by CANTA, however, much guidance and handholding is required for emerging NTAs. Strategies should be put in place to assist these member states in meeting the standards, and to benchmark their programmes internationally using a variety of external, national and regional assessment and certification agencies, which carry the confidence of both the public and their profession. These arrangements would provide training institutions with reliable and valid measures of key indicators of quality in the system including but not limited to: (1) the development of training outcomes, (2) training delivery, (3) assessment, (4) certification and (5) adherence to best practice.

A Monitoring System

CANTA, as the regional body, should have in place a system for monitoring NTAs that have the authority to award CVQs. This could possibly be in the form of a panel of independent professionals and experts, who can offer advice and feedback to the NTAs on a timely basis. This panel could act as "external inspectors", collecting information through observations and review of documents for feedback and improvement. The external audit is used primarily to check the internal evaluation against established CANTA standards. To assist NTAs even further, a "dry run" should be performed to familiarize training entities with the requirements prior to the first external audit. This approach allows the transfer of knowledge to enable emerging NTAs to develop at a pace that is in keeping with their own capacity and capability.

A Measurement Tool

Finally, a measurement tool should be developed and used to rank TVET institutions across the region. It is now standard practice in many countries to use performance indicators in the process of accreditation, evaluation and ranking of TVET institutions. With institutions facing increased competition for diminishing resources and with stakeholders demanding greater accountability, the use of performance indicators should be considered by CANTA to measure and rate NTAs and training entities. Simple, easy to use, evaluative tools with indicators can be used as part of a self-evaluation process to answer the question: "How well are we doing?" This would inform development planning and provide practical answers to the question: "What should we do to improve ourselves?" Using the ten United Nations Educational, Science and Cultural Organisation categories identified above as a guide, an assessment form can be developed with a scale using the following levels for each indicator (see example in appendix 8.3):

- Must improve as a priority
- Satisfactory but needs to improve

- Good, meets expectations
- Excellent, exceeds expectations

NTAs awarded a minimum score would then be authorized to award the CVQ. This measurement tool will allow for NTAs to identify where their weaknesses lie and to assess where they are in relation to other similar regional entities. The necessary strategies should then be put in place to improve on the weaknesses, drawing on the strengths of the other NTAs in the region.

Conclusion

Quality assurance is a powerful means to improve the effectiveness of TVET and to overcome the negative image of the sector. Given the urgent need for the region to increase its competitiveness and productivity through training and certification, countries have to be fully committed to making TVET a priority and giving the necessary attention to quality assurance in TVET. Self-evaluations and planning for development are key aspects of quality assurance systems. However, these processes are not sufficient for ensuring improvement. They need to be part of a fully fledged quality assurance system in which CANTA must take the lead. Much more handholding, technical guidance and information sharing are required by emerging NTAs. The use of short-term interventions by CANTA can create the conditions and provide the necessary support to empower emerging NTAs to develop a better understanding of the requirements. The proposed flexible compliance model is a start in improving the information and understanding in sharing best practices and building a continuously improving quality assurance system for TVET throughout the region.

Appendix 8.1. Regional vocational qualifications framework

Level of programme/Type	Orientation and purpose	Credits	Entry requirements	Occupational competence	Academic competence
Level 1/certificate	Completion of a preparatory programme leading to further study in a given academic or vocational area or entry qualification for a particular occupation	Minimum 10 credits	To be determined by the local training institution	Semi-skilled, entry level, supervised worker	Grade 10
Level 2/certificate	To prepare a skilled independent worker who is capable of study at the next level (post-secondary)	Minimum 20 credits	Grade 11 or equivalent	Skilled worker, unsupervised worker	Grade 11
Level 3/diploma and associate degree	A post-secondary qualification emphasizing the acquisition of knowledge, skills and attitudes (behavioural competencies) to function at the technician/supervisory level and pursue studies at a higher level	Diploma: minimum 50 credits; associate degree: minimum 60 credits	4 CXCs, Level 2 certification or equivalent	Technician, supervisory	Associate degree entry to bachelor's degree programme with or without advanced standing
Level 4/bachelor's degree	Denoting the acquisition of an academic, vocational, professional qualification in creating, designing and maintaining systems based on professional expertise	Minimum 120 credits	5 CXCs, Level 3 certification or equivalent	Competence that involves the application of knowledge in a broad range of complex, technical or professional work activities performed in a wide range of contexts. This includes master craftsman, technologist, advanced instructor, manager, entrepreneur	
Level 5/post-graduate/advanced professional	Denoting the acquisition of advanced professional post-graduate competence in specialized field of study or occupation		Level 4 certification or equivalent	Competence that involves the application of a range of fundamental principles at the level of chartered, advanced professional and senior management occupations.	

Appendix 8.2. Certification chart

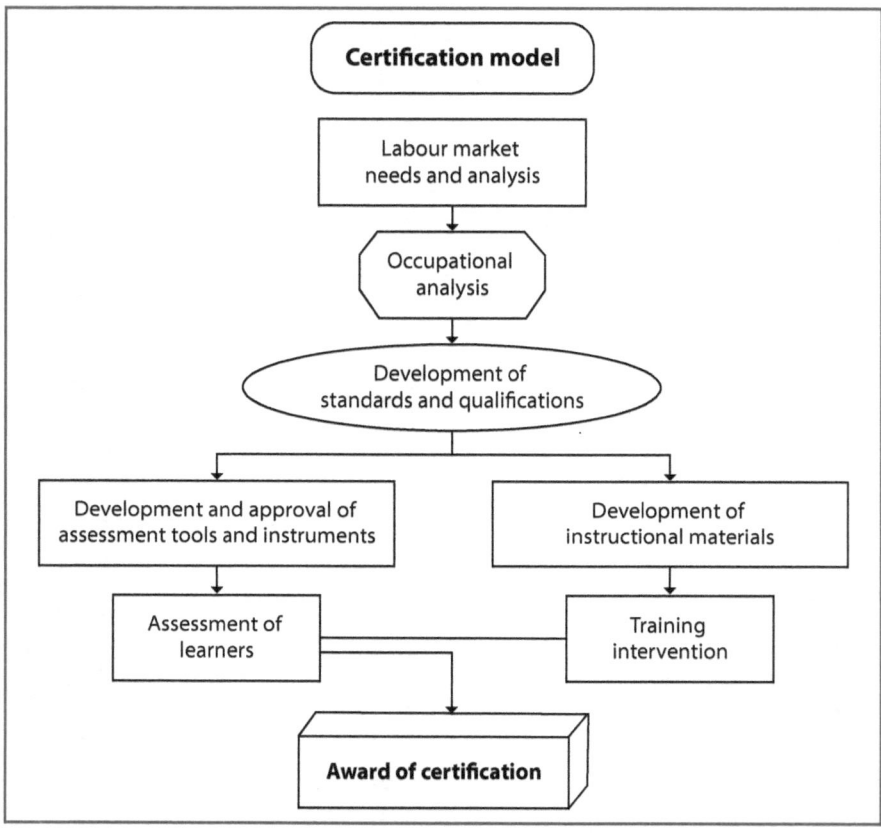

Appendix 8.3 Self-assessment tool

No.	Indicator	Must improve as a priority (1)	Satisfactory but needs to improve (2)	Good, meets expectations (3)	Excellent, exceeds expectations (4)
1	The use of regional assessment standards				
2	Competence of teachers and trainers				
3	Competence of TVET institutional managers				
4	Personal appraisal systems and identification of training needs of teachers and trainers				
5	Internal quality assurance and self-assessment undertaken by TVET providers				
6	Guidance and support services for learners/trainees				
7	Learning and teaching equipment and resources				
8	Reliable and timely data about TVET activity and outputs				
9	Collaboration and partnerships with employers and other stakeholders				
10	Links with parents and others who might influence the decisions of potential learners				

References

Bowen-Clewley, L., K. Cooper and R. Grannall. 2010. "A Comparison of Quality Management Approaches for the Training and Vocational Sector in Seven Countries". Paper presented at the thirty-sixth annual conference of the International Association for Educational Assessment, Bangkok. http://www.iaea.info/documents/paper_4d2287df.pdf.

CANTA. 2006. *CARICOM Process for Workforce Training, Assessment and Certification*. Kingston: CANTA Secretariat.

———. 2009. *CANTA Quality Assurance Guidelines*. Kingston: NCTVET, HEART-NTA.

———. 2012. *CARICOM Regional Technical and Vocational Education and Training (TVET) Strategy for Workforce Development and Economic Competitiveness*. Port of Spain: C-EFE, Trinidad.

Dunn-Smith, P. 2012. "Whither the CVQ, Challenges in Implementing the Caribbean Vocational Qualifications". Paper presented at the UWI-UNEVOC conference, Montego Bay, Jamaica.

ILO. 2010. "Employment Sector Employment". Working Paper No. 58, International Labour Organisation, Geneva, Switzerland.

UNESCO. 2010. *Guidelines for TVET Policy Review*. Paris: UNESCO.

UNESCO-UNEVOC. 2013. *Quality Assurance in TVET*. http://www.unevoc.unesco.org/.

3 | Internal Issues

9 | A Summative Evaluation Model for Strategic Planning in Tertiary Education

John Gedeon

There is no other single process that demands as much planning, staff hours, finance, resources, careful integration, monitoring and reporting, nor has a more far-reaching and strategic impact, than the strategic planning process. Indeed, the reputation and future of the institution rests on this single but involved process. Yet, after devoting that amount of effort, it is indeed surprising that few call for a thorough review to see if it was money and time well spent – or just a grand ritual performed every five years. The anecdotal evidence that a summative evaluation is well worth the cost shows that many strategic objectives and their related projects are never achieved by the end of the plan period and are rolled over into the new planning cycle. The vision is almost never achieved in its entirety. Even if strategic projects are achieved, are they having the intended impact on operations? The incredible number of variables in operation makes this far from a textbook procedure. Constant learning of what is working or not working with real-time feedback is required. In most organizations, it is no one's full-time job – not even university planners – who typically work on several other projects and research at the same time. There is an assumption that staff and management, with little or no background in organizational behaviour and management, strategic planning, or project management, can effectively develop and execute a very complex set of activities, when they are struggling even in making their everyday processes meet basic customer requirements. Learning is a key – whether strategic or operational activities are the focus. The summative evaluation – if done correctly – can provide the answers to keep from repeating the same costly mistakes over and over. All internal stakeholders want to know how they performed, as well as the institution as a whole. External stakeholders, especially funders and partners, want to have confidence in their investment.

Determining the quality of any system demands that systems be monitored periodically to identify whether objectives or standards are being met that will ultimately translate into stakeholder satisfaction. When it comes to the evaluation of the strategic planning activities and results, many universities focus on the *formative* methods by conducting periodic reviews of progress (mostly annual) during the standard five-year-plan period to make real-time adjustments. This chapter focuses on the *summative* evaluation process conducted at the end-of-plan period, which has largely been neglected in the literature. The model presented operates on the principle that the institution must first determine the "right things to do" (outcomes) and then "do them right" (outputs). At each stage, assumptions are made for both external and internal conditions and logic – many times unconsciously and many times incorrectly. This concern is encapsulated in the classic joke: "the operation was a success (output) – but the patient died (outcome)".

A comprehensive summative evaluation would have the following objectives:

1. Historically document the major activities and events in the strategic planning cycle (with references to more detailed information).
2. Determine the impact of the strategic activities on both the critical stakeholders and the institution itself, that is, its effectiveness.
3. Review the process of how the strategic plan was researched, created, executed, monitored, updated, adjusted and reported.
4. Determine the return on investment (were the results/benefits worth the effort expended).

Literature Review

A literature review was conducted to identify summative evaluation models for a comprehensive five-year strategic plan review. The research revealed most models focused on formative aspects, but there were two significant summative models discovered – one an elaboration of the other. In 2004, the Kellogg Foundation introduced a linear *Basic Logic Model*, shown in figure 9.1, comprising five sequential boxes ranging from "input" to "impact", which resembles a "systems model" but lacking any type of feedback loops. Dalrymple (2007) found this model to be too simplistic and created a *Comprehensive Logic Model* also shown in figure 9.1 (overlaid in *italics*).

Argyris (1977) introduced the concept of double-loop learning. Any system is concerned with basically two processes: its external relationship to its environment

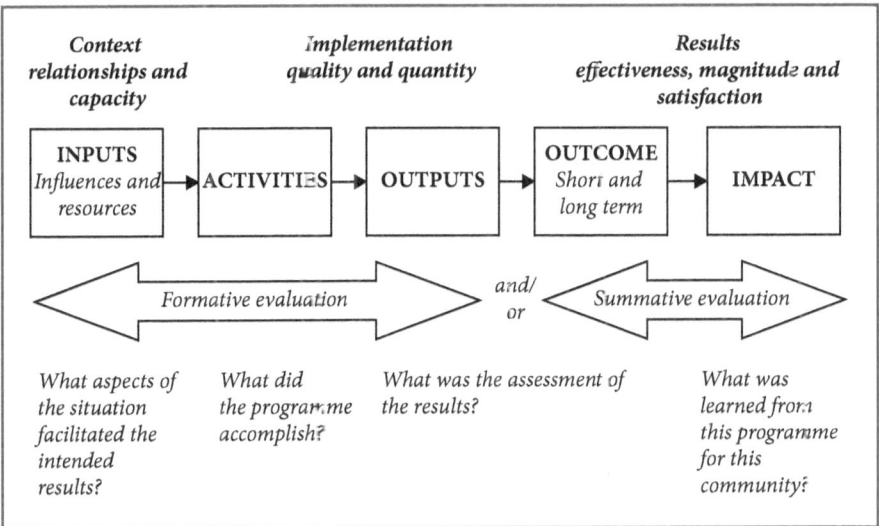

Figure 9.1. *Basic Logic Model* (uppercase) and *Comprehensive Logic Model* (elaboration in italics)

Sources: W. K. Kellogg Foundation (2004) found in Dalrymple (2007).

and its internal productive activities that create value. Systems cannot survive (that is, keep their identities) in a changing environment without adapting their internal processes to external realities. To adapt, a system must learn and all learning depends on feedback. Because the *Comprehensive Logic Model* was not itself comprehensive, it was enhanced by utilizing general systems theory to create a systems model employing double-loop learning feedback. Figure 9.2 lays out the generic model, which will be customized specifically to accommodate a summative evaluation of a strategic plan, while still incorporating the formative evaluations loop.

Double-Loop Systems Performance Model

The cycle in figure 9.2 starts on the left where the model interrogates and gathers data about its relationship with its environment, which is comprised of both human (stakeholders) and natural forces or conditions. Systems survive in a two-way process of needs fulfilment. The system needs resources from its environment and in turn fulfils the needs of these providers in some type of transaction. The system's identity is its mission, that is, why does it exist, what is its purpose? The mission identifies the system's key stakeholders and how they are to be served. So the "right things" to do is to meet the needs of its stakeholders and secure resources, while

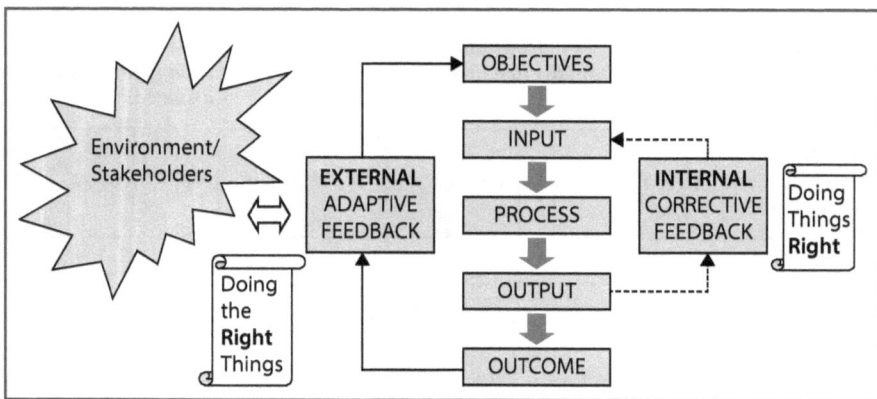

Figure 9.2. Double-loop learning systems performance model

protecting itself against any forces that would disrupt or distract from its mission (for example, competitors or disasters). The objectives answer the question: Given the nature of the environment and stakeholders and their needs, the system's direction and capacity, what should it strive to achieve?

A system is valued or relevant because it has the capability to convert inputs into outputs that are of value to stakeholders (mainly customers). Inputs are the resources, materials, competencies, information, facilities, equipment, money and other components that are required to create outputs by being organized into a process, which is a series of value-adding activities. The output must meet both quantity and quality standards (including cost) on schedule and when it does, then "things are done right". This is achieved through the internal corrective feedback loop, which detects variations from standards and then attempts to determine causes – that when addressed – will decrease variation (improve quality). The concepts of *output* and *outcome* are often confused because they are both results. The output is direct product or service from a process or a project (in the latter, it is termed a *deliverable*). The output has no value unless it causes the desired outcome.

Applying the System Model for Strategic Planning Evaluation

To customize the double-loop systems model for a summative evaluation of a five-year strategic planning period, the model components were translated into the areas of the strategic planning cycle from a summative viewpoint (not necessarily in chronological order) as outlined in table 9.1.

A Summative Evaluation Model for Strategic Planning 153

Table 9.1. System components translated for the strategic planning cycle

System component	Equivalent strategic planning component
Environment/Stakeholder	• Environmental conditions and assumptions – starting and emerging • Threats and opportunities (from SWOT) • All stakeholders – customers, regulators, funders and competitors
Objectives	• Vision and mission • Strategic goals • Strategic objectives
Inputs	• Strengths and weakness (from SWOT) • Values • Human resources • Financial resources • Informational resources • Infrastructural resources
Process	• Strategic projects (execution)
Outputs	• Project performance (deliverables)
Internal corrective feedback	• Project-level corrections ("doing things right")
Outcomes	• Process performance ("doing the right things")
External adaptive feedback	• Annual strategic plan retreat reviews • Five-year reviews

Summative Evaluation Questions and Data

The foregoing headings in table 9.1 are elaborated in tables 9.2–9.8, where they are listed alongside the major system model components. For each component, generic evaluation questions were generated by the author. The list is illustrative and not exhaustive, so evaluators can add or modify questions to make them pertinent to their situation. These questions can be qualitative or quantitative in nature. As a general rule, those components dealing with process (how) are qualitative and those with results (what) are quantitative.

An additional column labelled "Data sources" is provided to assist evaluators in identifying possible places to look for answers for the questions. If the source is a person (usually a job type or role played), one may want to conduct an interview; if a few similar persons are involved, a focus group may be most efficient; or if many people are involved, some type of survey or questionnaire may be appropriate. Be sure to collect all survey items for a given stakeholder from all seven tables when

Table 9.2. Strategic plan summative evaluation questions: Environment

System model component	Corresponding SP components	Evaluation questions	Data sources
Environment	Environmental conditions and assumptions – starting and emerging	• What were the significant environmental conditions during the SP creation?	SP/Public data
		• What emerging environmental conditions had a material impact on SP direction or execution?	Annual review meetings and reports/public data
		• What assumptions were made about the environment and stakeholders?	SP/planners
		• Were these assumptions comprehensive and true and did they hold for the entire SP period?	SP/public data/planners
	Threats and opportunities	• Were the threats and opportunities correctly identified?	SWOT analysis
		• Were threats addressed and opportunities seized?	SP
	Stakeholders – customers (students, engaged sectors and governments)	• Were all critical customers and their numbers identified?	SP/EMT and planners/public data
		• Were their current and future needs assessed?	SP/planners/public data
		• Were these needs translated into improved organizational systems, programmes, products and services?	SOs and projects

Stakeholders – funders	• Were all significant funders identified? • Were non-traditional funders identified?	SP/SPT/planners
	• Did pledged funding match actual funding?	Bursary reports
	• Were funders properly managed? • How were shortfalls addressed?	Bursary
Stakeholders – regulators	• Were all regulators identified? • Were all regulations complied with?	Legal and affected units
	• Did any new regulations create problems and how were these addressed; what was their impact on operations?	Legal dept/affected units
Stakeholders – competition	• Were all competitors identified?	SP and SPT
	• What elements of the SP addressed competition?	SP and SPT
	• Were the strategies successful?	KPIs
All stakeholders	• Were they satisfied with the way the SP was developed and executed? • Were they satisfied with the outcomes or results?	SP results and KPIs/all stakeholders

Notes: Abbreviations are in alphabetical order. EMT, Executive management team; HOD, head of department; KPI, key performance indicator; PI, performance indicator; Rx, research; SG, strategic goal; SO, strategic objective; SP, strategic plan; SPT, strategic planning teams ("planners" refer to full-time professionals); SWOT, strengths–weaknesses–opportunities–threats; "/" is used to separate major sources of data. (These abbreviations will be used in the subsequent tables, hence these notes will not be repeated.)

designing it. When internal documents are sourced they can reviewed for accuracy and comprehensiveness. Reports are especially important, but they are often issued many months after the report period covered and usually in an "achievements-only" format. "Progress" reports herein generally refer to the performance of strategic projects (initiatives), while "annual" reports are the formative summaries of many aspects of the strategic activities and their impact used to make real-time adjustments to strategy execution activities. At times, information external to the institution must be researched, especially to monitor the environment or key stakeholders. Many times the plan's creators may not have done a comprehensive job in this area.

At the beginning of the new strategic planning cycle, which could be the last year of the current plan, the planners need to start the planning process so that there will be a smooth transition to the new plan. The new plan will both roll over what has not been completed in the old plan and be updated to address new and anticipated conditions and intentions. Formative lessons learned will play a significant role in the design of the new strategic planning process. There are many approaches for scanning and assessing environmental conditions, but in general they examine any force that can affect operations and those who are impacted by operations intentionally (customers) or unintentionally (protest/interest groups, the natural environment and so on); refer to table 9.2. Because the environment is constantly changing, the initial scan only provides a snapshot of conditions used to formulate the plan. Emerging conditions can have a significant impact on operations requiring either adjustments in intention or execution. Monitoring, interpreting and reacting to these conditions has been called *strategic thinking*: What is going on? How could it impact operations? What coherent response is required? This is the heart of strategy.

SWOT

The environment can be seen as containing friendly opportunities (situations, resources, support and so on) and unfriendly threats (competitors, strikes, substitute products and so on). Were favourable conditions exploited and threatening conditions addressed through risk avoidance or mitigation? Risk management models provide approaches for assessing and addressing those negative conditions that one cannot directly control. SWOT (strengths–weaknesses–opportunities–threats) analyses are often used to generate strategic objectives (which will be addressed in the next section).

The balance of the environment area examines make-or-break stakeholders (not addressed in SWOT). Strategy is about better serving mission-specified stakeholders while negotiating other stakeholder interests for the best overall outcome. The primary stakeholders are the customers and – translated into the university context – they are the students, research sponsors and recipients of consulting services or intellectual capital in general. If their needs are not met in an affordable and timely manner, the very existence of the institution is threatened. The next most significant set of stakeholders are funders, especially where higher education is seen as a "public good" and tuition does not cover all expenses. Without sufficient resources, the institution will be limited in what it can offer or improve. Regulators can dictate the behaviour of the university in certain sectors, which has grown into the "compliance industry", where consultants advise institutions on the challenge of managing operations with many constraints and conditions that must be legally met. While each regulation may seek to assure a level of quality or forestall the customer being taken advantage of, they all have cost implications for operations. Competitors, while seen to be threats, are also stimulants to improvement as they force the universities to rethink their products and the stakeholders' experience.

Mission and Vision

The operational environment presents the *realities* that the university must engage in order to realize its mission – its landscape. The second system component ("Objectives", see table 9.3) then must specify the organizational intentions on a number of levels. How will it journey through this landscape to create value while sustaining and protecting itself? Over the years, its mission can evolve as the organization learns from its historical interactions. It may have had a simplistic or incomplete notion of who its stakeholders were or how exactly they added value. Whatever type of creature the organization was it is always lagging behind what it needs to be in order to be effective by its own measure – hence the introduction of the *vision* concept. How does the institution need to perform (read: capabilities) to serve its primary stakeholders? A vision or mission statement may not provide employees with a sufficient picture so often enterprises elaborate on these statements to ensure that they are being interpreted consistently.

Once mission and vision are clearly articulated, they create the framework for what must be achieved – the rest is strategy. The strategic goals and objectives are usually specified to focus energy in those critical areas that will make a difference. This can be conceptualized as "imperatives versus issues". Imperatives are critical areas that need to be addressed to achieve mission/vision and issues are those things that stand in the way of that. Goals are broad statements of intention that are almost eternal (for example, "improve the teaching and learning process"), and, therefore, have no deadline, while objectives are specific, time-bound strategies that will collectively move the institution towards its goals. The main problems here are not being able to identify or focus on the important drivers and trying to improve everything at once. Unconscious forces make one believe that one can process and handle much more than is realistic even in a collective setting. While strictly

Table 9.3. Strategic plan summative evaluation questions: Objectives

System model component	Corresponding SP components	Evaluation questions	Data sources
Objectives	Vision	• Was the vision realistic for a five-year SP given the resources, culture and SWOT?	EMT
		• Was it written in a way that it could be understood by everyone? • Was it articulated in quantitative ways?	SP/Quantitative standards/employees
	Mission	• Did it realistically describe why the institution exists?	Mission/history
	Strategic goals (general areas for improvement)	• Were goals an elaboration of the vision statement?	Vision and SGs
		• Did goals also address issues that stand in the way of vision achievement?	Vision and SWOT
		• Did goals focus a few critical areas or did they try to correct everything?	Compare with other universities
		• Who had oversight responsibilities for each goal area?	EMT
	Strategic objectives (specific areas for improvement)	• Were the SOs both necessary and sufficient to achieve the goals?	SGs, SOs and strategy maps
		• Was it clear to those developing projects what was called for in the SO (that is, interpreting them correctly)?	Project managers and team members
		• Were the SOs mapped to see their logical relationships? What was found wanting?	Strategy maps/EMT, SPT and planners
		• Were there enough resources to apply to all SOs?	Bursary/project managers
		• Were the time frames of the SOs realistic?	EMT, project managers and planners
		• How many SOs were achieved by their deadline?	SP reports
		• How many SOs were still not achieved by the end of the five-year period?	SP reports/project managers and planners

speaking strategic objectives are intentions and answer the question "what", they often directly state or imply strategy and answer "how".

Balanced Scorecard Method

The foregoing leads to a discussion of what type of strategic planning model is being used. While most contain the same elements, they offer differing procedures for creating and executing them. A popular framework is the *balanced scorecard*, which started off as measurement methodology in the early 1990s and has evolved into a full strategic management system. Whatever approach is selected, it should be customized for both the local culture and specific type of institution. The summative evaluation would attempt to determine the degree of fidelity to that (customized) model and expose its shortcomings. One of the most difficult aspects of realizing strategic objectives is that they cut across many organization functions, so no one individual has total control or knowledge of their workings. Systems theory is clear that changing any part of a system has implications for the other parts. The balanced scorecard approach utilizes strategy mapping in order to visualize the logic of the relationships between strategic objectives. This is a "paper test" to look for cause-and-effect, sufficiency, sequence, redundancy, conflicts and gaps in strategy – does it hang together conceptually? The strategy map is fraught with assumptions about how the strategies will work and interact with each other to create outcomes.

The next three system-model sections – Input > Process > Output – are the engine room that delivers the goods with the chief concern being "doing things right". Instead of speaking directly about routine operational processes, the summative model limits the discussion to developmental projects often called *initiatives*. Strategic objectives are realized by necessary and sufficient projects executed in the optimal sequence. The target of any developmental project is to improve a routine operational process (faster, better, cheaper, expand capacity, or provide a new capability). The output of a project (*deliverable*) has no value until it is "installed" into an operational process to improve it. Figure 9.3 illustrates this logic. The teaching

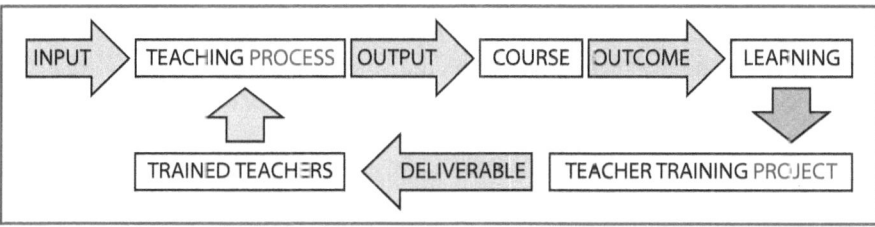

Figure 9.3. Project–process improvement relationship

and learning process produces an output called a course but the outcome sought is really student learning. In this example, the student achievement in a given programme was noted to be generally low so a project was formulated to train teachers (the deliverable) but trained teachers are of no value until they are operationalized (installed) back into the teaching process (that is, teaching).

To achieve results, there must be a capacity to achieve them and this speaks to the "Inputs" component of the model (see table 9.4). The first area of examination comes from a SWOT analysis that identifies, among other things, the internal strengths and weaknesses. There are various organizational assessment methodologies that can detect them. One rich source is the employee satisfaction or engagement survey. Staff can quickly put their fingers on shortcomings, as they must deal with the system on a daily basis. Strengths and weaknesses also speak to strategic behaviour. Is the organization exploiting its strengths (seizing opportunities) because of its competitive advantage in an area of its operations and is it protecting itself where its weaknesses expose it to risk?

"Values" are considered an input as they attempt to govern behaviour and can be classified as "eternal" (integrity, respect and similar concepts) or environment-specific like innovation or agility. Innovation was not valued during the Dark Ages – stability and conformity was. The environment provides clues about how to behave successfully. The evaluation must determine if the most important values were selected and if they were evident in management decisions and staff behaviour.

Employees can be a significant input depending on the way they are perceived. Traditional management theory views them as "labour" or a "means of production", which reduces them to machines that happen to be human. Strategic HR management views humans as the most important ingredient in productivity and quality. Strategic changes often require significant changes for staff. To what degree were they involved in the problem identification and planning stages to create ownership? It can be argued that, influenced by the colonial system, many times indigenous leaders (the elites) are deeply autocratic and paternalist; they often believe they know what is best and impose it on those of lower rank without their input thus creating resistance to change and even alienation. This is the point where true leadership can be ascertained. Leadership – as opposed to management – is where followers *voluntarily* follow the leader. Leadership Trait Theory (Kirkpatrick and Locke 1991) posits that one of the most important traits for a leader is *honesty* and when followers do not trust their leaders they only comply because they are being paid to do so. A high-performance organization is impossible without true leaders. "Buy-in" is not just into the programme but the leader, too.

Financial resources are required to fund projects that require out-of-pocket expenses, especially capital-intensive projects involving infrastructure works or large equipment purchases. In other cases, projects just require intellectual capital, but funds do not necessarily have to come from governments or even tuition. Some

Table 9.4. Strategic plan summative evaluation questions: Inputs

System model component	Corresponding SP components	Evaluation questions	Data sources
Inputs	Strengths and weaknesses	• Were all strengths and weaknesses identified correctly?	SWOT analysis/EMT and SPT
		• What strategies were used to address weaknesses and were they successful?	SP/EMT and SPT
		• Did the institution leverage its strengths to advance towards the vision? How?	EMT and SPT
	Values	• Were the values selected maintenance or developmental in nature?	SP values
		• What mechanisms were put in place to ensure that values guided decision making and general behaviour?	EMT and SPT
		• Were the values necessary and sufficient to guide behaviour?	EMT
	Human resources	• Were employees involved in the planning and execution of the SP?	SPT and project team members
		• How did management communicate the SP? • Was there general buy-in at all levels of the organization?	Marketing plan and reports/staff
		• Did the management have the requisite leadership and management skills to develop and execute the SP?	Management and consultants
		• Were project technical skills available in-house or were external experts required? • Was anyone trained in project planning or management? • Were there enough staff hours to complete all the assigned projects?	Project managers, planners and HR

(*Table 9.4 continues*)

Table 9.4. Strategic plan summative evaluation questions: Inputs (*continued*)

System model component	Corresponding SP components	Evaluation questions	Data sources
Inputs	Financial resources	• Were all projects properly costed?	Planning documents/project managers and bursary
		• How well did the budgeting cycle mesh with the demands of the SP activities?	Bursary
		• What happened to projects with insufficient financial resources?	Project managers and bursary
	Informational resources	• Most projects require data, was it available?	Project managers
		• What was done when data was not available for projects or decision making? How did it affect the quality of decisions or projects?	Project managers and ICT
	Infrastructural resources	• Were there sufficient physical resources (facilities, equipment, tools, materials, and so on) available to execute the projects?	Project managers and EMT/progress reports

can be self- or partner-funded or funded by diverting funds from some area of operations where a cost-saving through efficiency improvements has been realized. To what extent were projects properly planned so their true costs could be ascertained? To what degree is the budget system and cycle friendly to strategic project-based expenditures, as in many cases it is very recurrent-expense oriented?

Besides finance, one may argue that information is the lifeblood of any organization. Universities do not create physical products but essentially convert information in one form to another, hence the importance of information and communications technology systems. A major complaint of all information workers is access to accurate data in order to accomplish their tasks. Data may be in a paper-based form not published for internal circulation and a lot of time can be spent trying to find it. Often those who have the data tend to guard it as if were a state secret. Often data is over a year old when published, and it becomes historical rather than real-time. Real-time data can be acted on to address problems before they can cause maximum damage. Of course, the physical infrastructural resources must be available, too.

The "Process" component (see table 9.5) can be defined the same way, whether one is referring to projects (temporal) or processes (routine); projects being a series of activities that convert inputs into outputs. In the evaluation model, again, projects are the focus. Projects must be coordinated and tasks done in the correct sequence. When new (unplanned) projects were found necessary to introduce, how was this done and how were existing projects treated? At the end of a plan cycle, how many projects were still in the pipeline and what were the reasons for deadlines being missed?

In the "Outputs and feedback" component (see table 9.6), it must be determined if the deliverables of each project were done by the deadline, according to the quantity and quality attributes, and at what final cost. The better the planning, the more accurate will be the project duration and costs estimates, but as a rule these are almost always understated. This information is picked up in the first feedback loop, which is internal and can be termed as "corrective" feedback. It would specify criteria and mechanisms for project performance or progress reporting including database tracking. How was this formative information analysed, evaluated and communicated? Did executive management have an overall sense of how things were progressing so they could make high-level adjustments or interventions where necessary? What types of strategic meetings or retreats were held to make these decisions?

In the early stages of the plan, the focus is almost exclusively on project performance for two reasons: (1) many projects will take a year or more to complete and (2) even after the deliverables (outputs) are utilized in the target process ("operationalized") they will take time to have an effect. This last point is called the "outcomes" (impact) component in the evaluation model. Figure 9.4 depicts the relationship of a project to its target process. It is only when this process improves (as a result

Table 9.5. Strategic plan summative evaluation questions: Process

System model component	Corresponding SP components	Evaluation questions	Data sources
Process	Strategic projects	• Were the projects both necessary and sufficient to achieve the strategic objectives? • Were projects prioritized, sequenced, coordinated and integrated? • What mechanism was used to ensure proper project planning and management? • Were there any incentives for performance or consequences for non-performance?	EMT, deans, HODs and project managers
		• How were projects tracked? • How were new projects introduced after the SP was finalized?	Project database/ planning office and project managers
		• How many projects were still in progress at the end of the five-year period?	Progress reports/ project managers

of the project) that the project can be said to be successful. In other words, "doing things right" refers to the quality of the project outputs and "doing the right thing" speaks to the service or product of a process (the outcome), which almost always means satisfying stakeholders' needs (at an economic cost that meets all compliance regulations).

In the "Outcomes" component (table 9.7), one would want to know if the deliverable was actually being utilized in the process. Many times the deliverable is completed or approved and that is where the activity stops. For example, a policy is written and approved but not distributed or enacted with procedures and documentation. If the deliverable is operationalized then some impact or improvement in the process should be noted, as indicated in some type of metric called a performance indicator. If it is a critical organization metric then the term KPI (key performance indicator) can be used. Selecting the best performance indicator is important and it should be one that directly measures the improvement; if the performance indicator is too broad, then other variables besides the project could account for fluctuations (up or down) making it difficult to determine if the deliverable was the "active ingredient". Many times, because of poor problem (cause) identification the wrong

Table 9.6. Strategic plan summative evaluation questions: Outputs and feedback

System model component	Corresponding SP components	Evaluation questions	Data sources
Outputs	Project performance	• Were project deliverables completed as scheduled? • Did the deliverables have the stated quality and quantity attributes? • What were the final costs of the projects?	Progress reports/project managers and bursary
		• How were project completions communicated?	Progress reports/project managers and planners
Internal corrective feedback	Project-level corrections	• Was a project database set up? • Who monitored project performance? • How was project status and performance reported? • When milestones or deliverables were not met how was this addressed? • Of what type and frequency were project review meetings?	EMT, deans, HODs, project managers and planning office

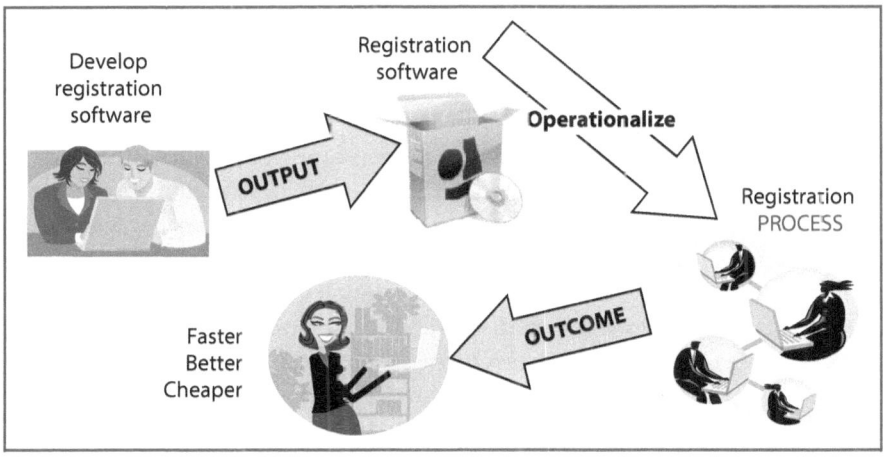

Figure 9.4. Project output and process outcome performance measurement points

Table 9.7. Strategic plan summative evaluation questions: Outcomes

System model component	Corresponding SP components	Evaluation questions	Data sources
Outcomes	Process performance	• Were completed project deliverables installed into the target process to improve their performance?	Progress reports/project managers and process owners
		• How was outcome (process performance) measured? • Did process performance improve? • Could other factors have an effect on these performance metrics? • How many outcome targets were actually met? • How were best practices that were discovered shared throughout the institution?	KPIs and PIs
		• When process performance did not meet targets, what was done?	EMT and process owners

solution (project) is used. In situations where process performance did not improve, what remedial action was taken?

Monitoring and Reporting

The monitoring and reporting process for both projects themselves and their impacts on routine processes are critical to the success of the plan. The number of variables and assumptions in operation make strategic execution anything but straightforward. In this model, it is called formative evaluation, which usually translates into annual reviews. It is a learning process because there are only guiding principles to be applied but no fixed answers, hence it is the equivalent to an experiment to see what works and to what degree. It is important to know what is going on in real time in order to make corrections as soon as possible. In the early part of the plan, executive management should be aware of project progress status in order to keep things on track and resolve any issues that are beyond the project teams' ability to control. As projects are completed, the spotlight can switch to process performance improvement (impact). A slow progress, at first, towards the target values should be evident if the cause-solution assumption is correct. The root cause may not have been identified and perhaps only symptoms are being treated. The solutions applied may be necessary but not sufficient. The sequence of implementing solutions may not be optimal or may be poorly timed. Any change in one part of a system eventually affects all of the system, and those intended and unintended consequences should be identified. Strategy maps can be used to check the cause-and-effect logic, necessary-and-sufficient requirement, integration from top to bottom, and sequence of the execution of activities at both the strategic objective and project levels. This is the equivalent of playing chess on a multi-tier board. The strategy map visualizes the relationships and allows errors to "jump out at" the planner in a way that their verbalization alone may fail to do.

Collecting data to monitor strategic operations has progressed from manual paperwork to digital representations of that data and, most recently, to software deliberately designed to collect, summarize, and display data to improve the analysis process. Some data can be collected by the system in databases as a matter of course and other data will need to be manually extracted. Depending on the type of project or performance measurement, the data would be summarized vertically from the unit to department, to the faculty (or school) and then to the campus level. No matter what system is being employed, someone (or team) must interpret and respond to what the data is telling them – it is difficult for this to be done by many people as a part-time responsibility given the magnitude of the investment in strategic activities. Kaplan and Norton (2005) recommend the setting up of a full-time office of strategy management, which can be one person in a small institution or a team of professionals in larger concerns. They suggest the following roles:

- *(Balanced) scorecard management* – Design and report on the balanced scorecard measures.
- *Organization alignment* – Ensure all business and support units are aligned with strategy.
- *Strategy reviews* – Shape the agenda for management strategy review and learning meetings.
- *Strategy planning* – Help the executive team formulate and adapt the strategy.
- *Strategy communication* – Communicate and educate employees about the strategy.
- *Initiative management* – Identify and oversee management of strategic initiatives (projects).
- *Planning and budgeting* – Link financial, human resources, information technology and marketing to strategy.
- *Workforce alignment* – Ensure all employees' goals, incentives and development plans link to strategy.

- *Best practice sharing* – Facilitate a process to identify and share best practices.

The author would like to add "conducting the summative evaluation of the strategic plan at the end of the five-year cycle". The last point, "best practice sharing", includes the lessons learned in process. The lessons would be converted into recommendations to be employed immediately where applicable or in the next round of the planning cycle.

The final component –"External adaptive feedback" – attempts to determine if the "right things were done" (see table 9.8). "Right" means that the stakeholders' needs were satisfied at a reasonable cost and by making significant strides towards the vision while minimizing risk. The environment is not to be visited once in every five years, but should be periodically scanned to detect any condition or trend that would necessitate a shift in strategy and can be considered an updated SWOT to include any significant changes in stakeholder needs. Strategic thinking is employed here to make adaptive changes in a dynamic environment. This information should be presented to executive management for consideration on an annual basis unless there is a compelling reason to address it sooner. It needs to be reported in a plan-versus-actual format instead of traditional "historical" method – which just lists achievements alone – and communicated to stakeholders.

What major issues prevented full achievement of the plan? Many of these will be found in the "W" of SWOT – internal weaknesses – especially in the plan process itself: how was the plan researched, created, organized, resourced, skilled, communicated, monitored, reported and reacted to? Sometimes lessons are *not* learned and the same mistakes are repeated. Here leadership becomes critical to lift the enterprise out of its culture-bound behaviour.

Organizing the Summative Evaluation

The summative evaluation report itself should cover two major areas, the strategic planning process and the results of that process, as expressed mostly in performance indicators. It cannot be completed until all units have reported and all data collection activities have been accomplished. This could take at least half a year after the close of the planning cycle, so the evaluation report itself may not be ready until a full year after the close. This creates an interesting problem as the planning of the new strategic plan should have begun by the beginning of fourth year of the previous plan. Hence, lessons learned in the summative evaluation would not be available in its entirety at that time. Therefore, it highlights the importance of formative evaluations in shaping future planning cycles.

While a lot of the data will be available from the system, much of it, especially when the data source is a person, will have to be collected from interviews, focus

Table 9.8. Strategic plan summative evaluation questions: Adaptive feedback

System model component	Corresponding SP components	Evaluation questions	Data sources
External adaptive feedback	Annual strategic plan retreat reviews	• How were stakeholders appraised of the SP execution both progress and problems? • Who scanned the environment for changes that would affect the SP? • How were these environmental changes communicated?	Reports and Ad messages/marketing, EMT and planning office
		• What schedule of formative review reports and meetings/retreats were held?	Reports and meeting documentation
		• Was reporting "historical" or performance reporting in nature?	Reports
		• What alterations to the SP came out of this formative data?	Reports/EMT and planning office
		• When projects did not have the intended effect on process performance, what was done? How was this documented?	Process owners, EMT and planning office
	Five-year reviews	• What types of final reports were required from those involved in managing or executing the SP?	Reports/EMT and planning office
		• To what degree was the vision achieved?	KPIs/EMT and planning office
		• What were the major issues that prevented full achievement of the SP?	Reports/EMT and planning office
		• How early was planning for the next SP started? • Did it use lessons learned from the previous SP?	Planning office records/planners
		• Was the return-on-investment worth the effort? (a sophisticated accounting system is required to perform this calculation)	Financial human resources utilized vs Value of results

groups, or surveys. This will be the most expensive and time-consuming aspect of researching the summative evaluation. Careful planning and coordination is required to get these tasks done quickly and at low cost.

If the institution has an office of strategy management or a planning office, it would have the responsibility to research, analyse and produce the report. The full-version report would go to all of management. Specialized and shorter versions could be prepared for staff and critical external stakeholders who have a form of investment in the institution – financial or otherwise. A sample table of contents for the final report could be:

- Executive Summary
- Introduction and Institutional Background
- Short History of Previous Strategic Plans
- Summative Evaluation Methodology
- Timeline of Major Events
- Creation of the Strategic Plan
- Strategic Plan Management (Executing and Reporting)
- Changes Made While the Plan Was Active
- Results: Outputs and Outcomes to Reach the Vision
- Lessons Learned and Recommendations

Conclusion

Besides being a guide to summative evaluation, this chapter can also serve to educate those who are new to the strategic planning process by providing an overview of all its components and how they are supposed to work together to propel the institution towards its vision.

References

Argyris, Chris. 1977. "Double-Loop Learning in Organizations". *Harvard Business Review* 55 (5): 115–25.

Dalrymple, Margaret. 2007. "The Value of Evaluation: A Case Study of Evaluating Strategic Planning Initiatives". PhD diss., Purdue University.

Kaplan, Robert, and David Norton. 2005. "Creating the Office of Strategy Management". Working Paper 05-071. http://www.hbs.edu/faculty/Publication%20Files/05-071.pdf.

Kirkpatrick, Shelley, and Edwin Locke. 1991. "Leadership: Do Traits Matter?" *Executive* 5 (2): 48–60.

10 | Internationalization, Cross-Border Tertiary Education and the Challenge for Quality
A Jamaican Perspective

DAMEON A. BLACK

Jamaicans, like other Caribbean people, regularly consume the services and products of international tertiary education providers. This pattern is motivated by perceptions of a higher, intrinsic quality of the international product; also, the view that such consumption leads to distinct personal, professional and vocational competitive advantages. Given the current economic climate, with the constrained employment and business activity, the comparatively high cost of that investment is justified in keeping with the perceived returns.

Quality is critical Here, it is understood as "fitness for purpose"; meaning, that which is being measured for its quality does what it is designed to do. The international education services and products delivered within Jamaica's borders are often marketed as being of the same quality as those offered to students on the providers' home campuses. These assurances are important, but there are concerns, given the importance of the issues of contextual relevance, interpretation and application. Quality cannot be reduced to a brand; while a brand may communicate quality, it is significantly more. Therefore, quality assurance mechanisms are necessary to safeguard the integrity and authenticity of Jamaican society and the commensurate investments of all stakeholders. This is where the role of external national accrediting agencies such as the University Council of Jamaica (UCJ) is most visible.

Globally, tertiary education entities are facing major challenges occasioned by resource constraints and the emergence of technology-facilitated delivery modes such as online and massive open online courses. The need for diversification of revenue streams and increasing inflows is paramount for these institutions, and with such market imperatives they will invest in contexts that are deemed to be rich with potential for significant returns on their investments. This leads to cross-border education becoming a significant income earner for institutions for developed nations and therefore an extractive experience for developing ones.

In the face of this market-driven expansion in tertiary education across borders, what are the challenges for the national external quality assurance agency? This chapter argues that in the face of the increasing international and cross-border nature of tertiary education, often as a means of generating new and increased revenue, a path has to be charted in Jamaica to facilitate creative engagements so that needed value is added, quality is assured, and the extractive process managed in the interests of the national good.

Internationalization

The concepts "internationalization" and "cross-border tertiary education" (CBTE) are related, yet sufficiently distinct and thus warrant separate definition. Knight (2006) views internationalization as "integrating an international, intercultural, and global dimension into the purpose, functions (teaching, research, service) and delivery of higher education" (cf. Stella and Gnanam 2005, 17). Internationalization may be achieved via campus-based activities in a domestic institution and does not require cross-border interaction. Critically, institutions engage in internationalization processes as strategic considerations dictate.

The use of the term internationalization in the first sense, as applied by Stella and Gnanam (2005) and Knight (2006), connotes development cooperation, whereby government to government or institution to institution projects are not defined with reference to the profit motive, but rather in relation to activities designed to benefit a nation or region, generally. One example of this is the project through which the United States Agency for International Development facilitated the development of the Mona School of Business; the Ohio State University was the international partner in this initiative. The Mona School of Business (now subsumed into the Mona School of Business and Management) is today a major business school with significant influence in corporate Jamaica and national life. The establishment of the G.C. Foster College of Physical Education in the 1970s is another example of development cooperation for national benefit. The Cuban government built the institution as part of its international development cooperation effort in the 1970s. Building capacity by developing institutions, training personnel in critical disciplines, funding and conducting collaborative research, and providing technical support are dimensions of the development cooperation approach.

The development cooperation perspective may be broadened to include emphases on revenue enhancement, while maintaining the altruistic orientation. Thus, while critical, profit and revenue are not the major drivers of such engagement. One example of this expanded perspective is the collaboration between the Caribbean Graduate School of Theology, in Kingston, Jamaica, and Bakke Graduate University (Seattle, Washington) to provide doctoral level study for students in the region.

Cross-Border Tertiary Education

The term "cross-border tertiary education" is used in keeping with the general appreciation that the qualifier "tertiary" is more inclusive than "higher" and much more reflective of the dynamic nature of the Jamaican tertiary education sub-sector. At the Mona Academic Conference on Tertiary Education in 2004, then minister of education Maxine Henry-Wilson, referred to a Government of Jamaica White Paper on education, issued in 2001 which defined tertiary education as "post-secondary education including higher education. Tertiary education no longer means university alone, as it . . . covers a wide range of education and tertiary opportunities . . . a range of formal tertiary training, the development of skills, and self-directed learning" (2005, 3).

The reality is that CBTE is substantially similar to the widely used cross-border higher education, so in this discussion the conceptualization and discussion of CBTE will include references to cross-border higher education. Cross-border tertiary education or cross-border higher education is thus understood as the movement of people (learners and staff), projects, providers, institutions and programmes (including curricula) across national or regional borders. The discussion will be focused on CBTE at the degree (undergraduate and graduate) level. The issues of sub-degree (vocational, diplomas) and professional certifications (for example, Association of Chartered Certified Accountants), though important, raise additional and different challenges that cannot be dealt with in this discussion.

Recognizing the increased commercialization of cross-border education at the tertiary level, UNESCO produced *Guidelines for the Quality Provision of Cross-Border Higher Education* in 2005. The publication has served as a critical reference tool for external quality assurance agencies and professionals in the effort to ensure that the provisions of the international providers conform to generally accepted standards for the teaching learning process at the highest levels of the education system.

The commercialization of tertiary education is facilitated by specific international trade agreements like the General Agreement on Trade in Services. General Agreement on Trade in Services defines four modes of supply or channels by which services may be traded: cross-border supply, consumption abroad, commercial presence and movement of natural persons. CBTE utilizes each of these channels. Within the Caribbean, cross-border supply is experienced when Jamaicans, St Lucians or Barbadians (for example) choose to study with the University of Phoenix or one of the other online providers. Consumption abroad is where people choose to leave their country to study in another. The decision by University of New Orleans or Nova Southeastern University to establish and operate offices in Jamaica at various points to deliver programmes and courses is an example of commercial presence. Movement of "natural persons" deals with educational professionals in one country travelling to another to offer educational services. Thus, visiting professors, who travel to Jamaica to deliver courses, or persons invited by the UCJ to sit on institutional assessment panels are participants in this fourth channel for trade in services.

Internationalization and Implications for Nation-Building and Social Cohesion

Chancellor of the University of the West Indies, Sir George Alleyne, in a discussion of the purpose of tertiary education, asserts that, beyond the production of new knowledge, human capital and technology, "the Jamaican higher educational system has to concern itself with the preservation of the culture, the values and some of the ancestral morés that are as essential for a people as their economic wealth" (2005, xviii). The role of tertiary education in providing such critical services for the communities of interest served by national (which should be understood as "public", given the chancellor's perspective) tertiary institutions is clear in Alleyne's list. The chancellor raised the issue against the background of the increasing role in the Caribbean of a number of private, for-profit institutions who see tertiary education exclusively as a private good. Alleyne is quite dismissive of such entities, adjudging them "profit-making degree-granting mills" (xxi). The chancellor is neither willing to depend on nor to trust such private institutions with the "value-affirming functions" that are so vital to building people and societies (xxi). Dr Muriel Howard, president of the American Association of State Colleges and Universities, while hailing from a different context, echoes Alleyne's concerns. She writes:

> The higher education-as-a-commodity mentality is increasingly omnipresent, with advertising saturating virtually every medium of our daily existence . . . Yes, possession of a diploma should be indicative of a graduate's qualified possession of certain intellectual capabilities – as well as enhanced creativity, flexibility, and analytical abilities. Yes, higher education institutions have an obligation to help students achieve those goals, to change with the times, and to adapt with technology. But if a university diploma is to have any meaning at all, it must continue to be a distinction that is earned, not a commodity to be bought, and change must be guided by those who understand teaching and learning, not buying and selling. (2013, I)

Cevallos (2013), while appreciating Howard's perspective, cautions that educators cannot act like Quixote and deny the reality of the landscape and the environment – an environment impacted by continuously expanding commercialization.

It is not possible to return to a state where this is not so; the globalized economy with its ruthless cut and thrust engagement does not allow this option.

The Role of Increased Demand

The contemporary growth in internationalization and CBTE is primarily driven by the explosion in demand for tertiary education, especially in developing countries where governments are challenged to satisfy same, given resource constraints (cf. Miller, 2005; Knight 2006; Leo-Rhynie 2005; Middlehurst and Fielden 2011). The level of demand for tertiary education is captured by Haddad (2003), who posits that if 10 per cent of the adult population in developing countries were to seek educational services at the tertiary level, this would represent an additional three hundred million persons actively participating in that vital sector.

Caribbean governments have also set targets for increased levels of critical cohorts being enrolled in tertiary education. In 1997, the Caribbean Community Heads of Government set a goal of 15 per cent of the eligible cohort of high school graduates throughout the region pursuing tertiary education by 2005. Foster-Allen notes that this goal was expanded in a draft strategic plan (developed in 2004–5) for tertiary education in Jamaica to 33.3 per cent of the eligible cohort by 2010 (2007, 123). Such targets place significant strain on resources within Jamaican public institutions. Importantly, such constraints facilitate the development of a market for CBTE and private (Jamaican) providers to achieve the targets and meet the demand (Henry-Wilson 2005, 9–10).

Traditional providers of tertiary education services in Jamaica might have unwittingly aided and abetted the increased demand for cross-border education. Anecdotally, such providers tended to be arrogant in the execution of their mandates as is seen by indifferent student services, constricted access, and inflexibility in scheduling and financing options for students. Additionally, their cost structures were inefficient. Henry-Wilson noted, at the above referenced Mona Academic Conference, that "compounding the problem of declining resources is the inefficient use of these resources. Higher education is to a great extent characterized by low student-staff ratios, under-utilized facilities, duplication of programme offerings, high dropout and repetition rates, and a very large share of the budget devoted to non-educational expenditures such as subsidized student accommodation, food and other services" (2005, 9). Taken together, these factors created perfect market opportunities and space for nimble entrants to occupy and make their own.

At the same time, there is no denying that there may be ideological imperatives at play in the increasing global extension of tertiary education from the developed to the developing world. Simon Lester (2013) seems to be rubbing his hands together gleefully at the thought of more, not less, competition for traditional providers. While more persons, globally, are securing degrees from foreign institutions, he is not pleased with the fact that tertiary education remains a predominantly national endeavour. For Lester, the onset of new online initiatives, such as massive open online courses, represent a further opportunity for transnational entities to be the dominant players in this sub-sector. To prevent nationalist sympathies from stymieing these initiatives, he argues that governments should resist protectionism and agree on new international rules to liberalize online education. How is the national tertiary education system to be sustained in the face of such a dominant perspective?

The Quality Imperative

Undoubtedly, in the face of ideological, commercial and demand/supply concerns, there are quality challenges associated with commercialized CBTE. Three challenges are highlighted from the Jamaican perspective. First, there is an encroachment on the ability of the state to secure policy objectives and establish credible and authentic regulatory frameworks. This has nothing to do with the integrity of current external quality assurance processes with the UCJ at the apex of the system. Rather, it acknowledges the fact that online entities, for example, can offer their products at will by technology fed right into the domestic space of students, irrespective of their geographical location. We are not only speaking about representatives of the private for-profit industry such as the University of Phoenix, but also venerable entities such as the University of London, where anyone may register for degrees and sit the exams at scheduled intervals at the Overseas Examinations Commission. Their degrees facilitate access; one such, the Bachelor of Laws, allows persons who pass the entrance exams to attend the Norman Manley Law School to qualify as a legal practitioner. Second, we have to deal with quality in the provision of goods and services of CBTE within Jamaica. There are fine examples of institution-to-institution partnerships that are characterized by quality as recognized through the process of accreditation. In 2013, the Association to Advance Collegiate Schools of Business lauded the collaboration between University College of the Caribbean and Florida International University in the delivery of the Master of Business Administration (Executive) in Jamaica. This programme is accredited by the UCJ. There is a long list of accredited programmes offered by CBTE providers in Jamaica, either by partnerships or through branch campuses. These providers include Central Connecticut State University (master's degrees in educational leadership and administration), Temple University (degrees in education with teachers colleges) and University College Birmingham (tourism programmes in collaboration with community colleges). Other cross-border providers operate *within* Jamaica but have chosen not to interface with the UCJ; these include University of Phoenix, Walden University and the panoply of British-based entities (via rdi). What are the quality and the cultural and contextual relevance of the curricula offered by the latter group? In this regard,

Leo-Rhynie (2005 referencing Contreras 2004) questions the quality of the education offered by Ross University's medical and veterinary schools in the Caribbean. She asks, "Ought we to tolerate 'brand name' universities offering the Caribbean 'watered down' versions of their home-based programmes?. . . Is there adequate technical capacity and support? Are tutors of acceptable quality?" (279). Her questions and emphases lie at the heart of the external quality assurance process, led by providers such as UCJ. Her questions are still relevant. Although she spoke with reference to medical institutions, the advent of the Caribbean Accreditation Authority for Education in Medicine and other Health Professions notwithstanding, the issue remains of ensuring that extra-regional tertiary institutions adhere to the standards of quality of regional external quality assurance agencies.

Third, CBTE providers *within* Jamaica tend to deliver market-oriented programmes in business administration, finance, education and information technology. Such programmes often require comparatively lower investments in infrastructure and systems and processes. And because this is so such programmes are delivered at price points that make it difficult for national entities to compete for students or to provide the add-ons that suggest high value-added. This is the extractive aspect of CBTE. With the perceived qualitative and competitive edge or the calculation by Jamaican consumers that such degrees assist their chance of assimilation when they migrate, Jamaican and Caribbean tertiary institutions lose the income from those students. This depletes the income available to Jamaican and Caribbean (not-for-profit) institutions to invest in vital areas – research, humanities, culture and performing arts, and science disciplines – reflecting the national developmental agenda. This severely challenges the fitness for purpose agenda of entities such as the College of Agriculture, Science and Education or of faculties within University of Technology (Jamaica), Northern Caribbean University or the University of the West Indies.

Tertiary education systems in Jamaica and other Caribbean nations have developed in the context of a global shift in paradigms from aid to trade. Within these systems institutional knowledge, learning and experience exist to enable the continued transformation of the tertiary sector to withstand the competition. Within Jamaica, the process of institutional strengthening is being advanced through the Jamaica Tertiary Education Commission. Jamaica Tertiary Education Commission is being designed as the body to lead and manage the strategic development and regulation of the tertiary sector within Jamaica. The pace of this development seems set to continue at a tedious pace.

Concluding Thoughts

Tertiary education faces a dynamic and heady future. National systems will be opened to increasing competition and commercialization. Critical questions remain,

"What is the role of higher education in today's society? What is the right balance between institutional autonomy and public accountability?" (Leo-Rhynie 2005, 279). The points made by Alleyne in advocating the continuation of government budgetary support are still valid: public tertiary education provision is important for social equity; to fund public goods such as research, preservation and transmission of cultural values; significant return on investment; to sustain genuinely national entities; and to ensure that institutions do not ignore their main functions in pursuit of entrepreneurial activities (2005). Society and the tertiary education sector continue to grapple with these questions. There is no need to fear the process wherein commercial and market-driven considerations are paramount. Given its inevitable advance, Caribbean nations need to study it, learn from it and co-opt it to serve their interests. In the contemporary climate, public institutions have to be more efficient as well as generate an increasing share of their revenue in the open market and via profit-oriented projects. Similarly, private institutions, consistently bereft of direct government funding, will have to be increasingly creative, innovative, and efficient in strategizing and using resources. The emphasis on the market is thus welcome if citizens benefit from educational processes that are defined by quality, characterized by relevance, allow for mobility, are transferrable in the global marketplace, and contribute to national and organizational development.

Haddad notes that in a competitive environment the "determining elements" are quality, efficiency, cost-effectiveness, relevance, convenience and prestige (2003, 6). There may be some discomfort with the latter two elements but they all play a part. Definitions for these elements are not provided by Haddad. This writer agrees with this approach as defining these elements should be the subject of various discussions throughout the system and in institutions in an effort to interpret and apply (perhaps even reconfigure) as per different contextual realities. These elements, and their discussion, contextualization and ownership (in whatever order of priority), constitute the path ahead for Jamaican (both private and public) tertiary institutions. If the fight is seen as one for survival, there will be significant losses; Jamaican institutions must be proactive in the process of developing agility, flexibility, efficiency and innovative approaches. At the outset of this discussion, reference was made to the need for the facilitation of "creative engagements". The primary message here is that Jamaican institutions must deem (and thus actively seek) the creation of partnerships and collaboration with CBTE providers as opportunities to build institutional capacity and for the generation and development of resources in a context of constrained national government expenditure and weak economic performance.

Further, with reference to the concerns posited by Alleyne (2005) and the quality issues highlighted by Leo-Rhynie (2005), the quixotic path must be rejected. Leo-Rhynie, Alleyne and others are not suggesting chasing windmills, but care to reject any and all arguments for the development and implementation of a regulatory framework that severely restricts the ability of CBTE providers to enter or access the Jamaican market. Any such approaches will negatively affect local higher education institutions. In the age of globalization with the rapid advances in information technologies, CBTE entities have virtually unhindered access to global markets. Thus, in this regard it is important to conclude the long overdue development and activation of the proposed regulatory framework within Jamaica – the Jamaica Tertiary Education Commission. Allied with this effort is the need to assure the place and role of the UCJ as the premier external quality assurance agency. These processes must lead to an environment in which both Jamaica Tertiary Education Commission and the UCJ do not act as barriers or bulwarks against, but rather as enablers of an environment in which Jamaicans (and internationals who consume the educational goods and services in Jamaica) are confident of the quality of the programmes on offer.

References

Alleyne, George. 2005. "Public Good and Personal Gain in Higher Education". In *Revisiting Tertiary Education Policy in Jamaica*, edited by Rheima Holding and Olivene Burke, xiv–xxxiii. Kingston: Ian Randle.

Cevallos, F.J. 2013. "Against the Windmills: The Commoditization of Higher Education". *Presidential Perspectives*, 2012–2013 Series 1.1–1.3. http://www.presidentialperspectives.org/pdf/2013/2013-Chapter-0-and-1Against-theWindmills-HE-Commodizitation-Cevallos.pdf.

Foster-Allen, Elaine. 2007. "Prospects for Tertiary Education in Jamaica". In *Higher Education: Caribbean Perspectives*, edited by Kenneth O. Hall and Rose Marie Cameron, 119–31. Kingston: Ian Randle.

Haddad, Wadi D. 2003. "Tertiary Education Today: Global Trends, Global Agendas, Global Constraints". Paper presented at the ICETE International Consultation for Theological Educators, High Wycombe, UK. August 19.

Henry-Wilson, Maxine. 2005. "Towards a Higher Education System for Jamaica: The Government Perspective". In *Revisiting Tertiary Education Policy in Jamaica*, edited by Rheima Holding and Olivene Burke, 3–10. Kingston: Ian Randle.

Howard, Muriel. 2013. Foreword to *Responding to the Commoditization of Higher Education*, by Aramark Higher Education, I–II. http://www.presidentialperspectives.org/pdf/2013/2013-Chapter-0-and-1Against-theWindmills-HE-Commodizitation-Cevallos.pdf.

Knight, Jane. 2006. "Higher Education Crossing Borders: A Guide to the Implications of the General Agreement on Trade in Services (GATS) for Cross-Border Education: A Report Prepared for the Commonwealth of Learning and UNESCO". COL/UNSECO. http://www.madrid2013.bolognaexperts.net/page/background-documents.

Leo-Rhynie, Elsa. 2005. "Diversity, Liberalization and Competition in Tertiary and Higher Education: Implications for Quality Assurance". In *Revisiting Tertiary Education Policy in Jamaica*, edited by Rheima Holding and Olivene Burke, 269–83. Kingston: Ian Randle.

Lester, Simon. 2013. "Liberalizing Cross-Border Trade in Higher Education: The Coming Revolution of Online Universities". *Policy Analysis* 720: 1–16. http://www.cao.org/publications/policy-analysis/liberalizing-cross-border-trade-higher-education-coming-revolution.

Middlehurst, R., and J. Fielden. 2011. "Private Providers in UK Higher Education: Some Policy Options". Higher Education Policy Institute. http://www.gov.uk/government/uploads/system/uploads/attachment_data/file/207128/bis-13-900-privately-funded-providers-of-higher-education-in-the-UK.pdf

Miller, Errol. 2005. "The University of the West Indies, Mona, and Tertiary Education in Jamaica". In *Revisiting Tertiary Education Policy in Jamaica*, edited by Rheima Holding and Olivene Burke, 60–103. Kingston: Ian Randle.

Stella, Antony, and A. Gnanam. 2005. "Cross-Border Higher Education in India: False Understandings and True Overestimates". GIME. http://www.gime.dufe.edu.cn/ieresearch/UserFiles/files/2011032219365870.pdf.

UNESCO. 2005. "Guidelines for Quality Provision in Cross-Border Higher Education". UNESCO. http://www.unesco.org/education/hed/guidelines.

11 | A Quality Scorecard Approach to Analysing Quality in Distance Online Education

PAMELA DOTTIN

The University of the West Indies (UWI) is a four campus institution, one of which is an Open Campus that was created in 2008. The Open Campus was created to expand the reach of the University and to provide access to non-campus countries and other underserved communities (*UWI Strategic Plan 2007–2012*). One of the methods that the Open Campus uses to expand the footprint of the university is the delivery of bachelor and master's programmes using online technologies.

Through its Quality Assurance Unit, the university conducts periodic evaluations of all its programmes. However, the evaluation mechanism has not been adapted to the different needs of the online environment of the Open Campus. Therefore, this study will discuss the applicability of Shelton's quality scorecard to the online provisions at the UWI. It will advance a thesis that the adoption and adaptation of that scorecard would better meet the needs of the online environment than what is currently being used for that modality by the university.

Quality, Quality Assurance, Quality Evaluation and Online Education

Traditionally, the concept of quality in higher education was restricted to a small, elite group of students who studied on a physical campus and attended classes in real time. For some, quality was seen to exist only in institutions that were expensive and exclusive (Daniel, Kanwar and Uvalic-Trumbric 2009). However, with the advent of mass education, there was a reconceptualization of what quality in higher education really meant. This debate was further deepened with the advancements in online education around the world. Cassey (2008) noted that the delivery of higher education in the online modality "holds greater promise and is subject to more suspicion than any other instructional mode in the 21st century" (45), perhaps, suggesting the need for the development of quality assurance mechanisms specifically for that environment.

Quality in traditional higher education is a difficult concept to define as it means different things to different people; for example, quality for a parent may mean value for money. Harvey and Green (1993) suggest that there are five broad concepts of quality: (1) exceptional quality; (2) consistency – zero deficits; (3) fitness for purpose; (4) value for money and (5) transformation. Quality assurance, on the other hand, is "a process of maintaining standards reliably and consistently by applying criteria of success in a course, programme or institution" (Mishra 2006, 100). Some authors agree that quality assurance consists of four components that must be assessed: (1) whether everyone in the institution takes responsibility for enhancing quality; (2) whether everyone in the institution has a responsibility for maintaining quality; (3) whether everyone in the institution understands, uses and feels ownership for maintaining and enhancing quality and (4) whether the management of the institution regularly check the validity of the system for checking quality (Frazer 1992, quoted in Mishra 2006).

Similar to quality, quality evaluation is difficult to define as its meaning is dependent on its use. Quality evaluation considers several different elements depending on who is using it and may include student feedback, curriculum, instructional design, technology, faculty qualifications and training (Meyer 2002). Quality evaluations, regardless of modality, must include mechanisms to continually monitor, evaluate and improve the quality of the provision. Many authors have proposed evaluation strategies for online education, including Lee and Dziuban (2002, quoted in Shelton 2010). The research indicates that quality in an online programme is mostly dependent on the assimilation of the evaluation processes into the normal day-to-day functioning of the programme and the institution. Benson (2003) explores quality in online education from a different perspective. She investigates how quality in online programmes is viewed by stakeholders. In research, she discovered that potential students, and other stakeholders, considered the following as indicators of quality in online education: accreditation, effective pedagogy and the institution's or programme's ability to overcome the stigma associated with online provisions (Shelton 2010). For this study, quality in an online programme is defined as "fitness for purpose", that is, the effective and efficient incorporation of evaluation processes into the planning, development and delivery of the online programme.

Quality Assurance and Online Education

The emergence of online higher education has also impacted on the field of quality assurance, both internal and external. It has led to the rethinking of quality assurance processes to meet the differing needs of the online environment and its students, as well as, to address many of the questions engendered by this modality. For example, is the quality of online programmes as good as face-to-face programmes? How should the quality of online programmes be evaluated?

This need to measure, assess and evaluate the quality of online education and its comparability to face-to-face programming has led to the development of many quality toolkits including Quality Matters, an industry-recognized quality seal for online programmes, developed by the University of Maryland; a rubric for online instruction developed by California State University, Chico; and the exemplary course rubric developed by Blackboard. Before these tools were developed, research conducted by the US-based Institute for Higher Education Policy (IHEP) in 1998 confirmed the need to increase the amount of research conducted into online programmes and the quality of those programmes. In 2000, the IHEP conducted a follow-up study which identified twenty-four quality indicators for online programmes. These indicators were grouped into seven overarching themes: institutional support, course development, teaching and learning, course structure, student support, faculty support, and evaluation and assessment. The IHEP further produced a set of characteristics whose presence would support each of the themes and would be an indicator of quality in the online provision.

In 2008, Dilbeck surveyed over two hundred community college administrators using the IHEP indicators to ascertain their perceptions of the quality indicators for online education. The Dilbeck study found that the twenty-four IHEP indicators were appropriate for the evaluation of online programmes. However, the tool which the IHEP used was not developed to foster quality enhancement. Campbell and Rozsnyai (2002) suggest that quality enhancement, also known as quality improvement, should be seen as a component of quality. This view is also shared by Vlăsceanu et al. (2007), who define quality enhancement as: "[F]ocusing on the continuous search for permanent improvement, stressing the responsibility of the higher education institution to make the best use of its institutional autonomy and freedom. Achieving quality is central to the academic ethos and to the idea that academics themselves know best what quality is" (72–73).

The search for an effective mechanism to evaluate the quality of the administration of online education programmes continued and it inspired Kaye Shelton to modify the IHEP's quality online indicators (for her PhD thesis submitted in 2010).

In her study, Shelton used the Delphi research method to ascertain the views of experts at forty-three institutions. From these findings, Shelton developed a quality scorecard for the evaluation of online programmes. Linstone and Turoff (2002) described the Delphi method as an interactive process used to collect anonymous input from experts. It collects the data using a multitude of collection techniques. The data collections are interspersed with feedback to the respondents. Shelton's study attempted to address not only quality assurance but sought to provide a mechanism for quality enhancement. The quality scorecard, which Shelton developed, uses a weighted four-point Likert scale and requires institutions to evaluate their performance against nine quality indicators. It also provides an interpretation of the final score and its implication(s) for the quality of the online programme.

The rest of this chapter will focus on using this tool to assess the current evaluation instrument used by the university. First, however, it is necessary to examine the evaluation mechanism that is currently employed by the university in order to identify the gaps that the Shelton scorecard could bridge. Importantly, it is necessary to take account of the structural, teaching and learning, learner demographics, and pedagogical differences that are a feature of the Open Campus; these must be reflected in any instrument used to assess the online programmes.

Peculiarities of UWI's Online System

The academic structure of the physical campuses consists of deans, faculty board, departments and faculty. The Open Campus, however, has an Academic Programming and Delivery Division, which consist of three departments: Programme Planning, Course Development and Programme Delivery. On the physical campuses, a department or a member of a department may propose a programme or be responsible for a programme. At the Open Campus, the development and delivery of a programme involves all three departments. The Open Campus has no faculty, so it recruits course writers to write its programmes and e-tutors and course co-ordinators to teach and deliver programmes. Indeed, the Open Campus was structured without faculty as the University believed that it could capitalize on the faculty members from its physical campuses for the development and delivery of its programmes.

In addition to this, there is the added requirement to have technical professionals troubleshoot and provide assistance to both learners and faculty. Yusuf-Khalil (2006), when evaluating a UWI online programme, noted that technological constraints hindered the use of discussions outside of the formal teleconference sessions. Although, the Open Campus now uses mainly Blackboard Collaborate for the delivery of its teaching, the scope for additional discussions outside these scheduled fora is still not readily available. Teaching staff play an important role in the quality of a programme and learner success as they create the environment that fosters or hinders learning. It is possible for facilitators to use the technology but not capitalize on its interactive potential; this potential creates a social constructivist learning environment, an environment that is viewed as strongly correlating with student success. Therefore, it is important that online facilitators and tutors are equipped with the necessary skills and attributes to adequately assist learners to succeed. According to Schrum and Hong (2002), "faculty members may know little about how to assist students in succeeding in [the online] environment and students may be ill prepared to tackle the new demands put upon them" (57).

Training online tutors is vital since the Open Campus recruits a number of its e-tutors/facilitators and course co-ordinators from the physical campuses of the University and many from outside the University; these persons often have little knowledge, experience or skills to teach in the online environment.

As important as orienting teaching staff to the use of new technologies is the challenge of transforming the traditional learner into an effective distance online learner. Yusuf-Khalil (2006), when reviewing a gender and development programme offered online by the UWI, noted with concern the inability of the e-tutors to create an effective student-centred learning environment "because of students' former orientation to learning that promoted the 'Banking Concept' of education" (Freire 1970, cited in Yusuf-Khalil 2006), which emerged as a major challenge for that particular programme. Yusuf-Khalil inferred that this dependence on the banking model was mainly due to the didactic-dependent nature of students in the Caribbean.

A further complication that must be addressed in the online environment and indeed at the Open Campus is the notion or concept of transactional distance. Transactional distance affects students in the distance education environment as there is a separation between the physical location of the tutors and learners. The students of the Open Campus come from seventeen anglophone Caribbean islands. Similarly, its tutors and course co-ordinators are recruited from the Caribbean and beyond. The separation of time and space between the tutors and students, plus their cultural differences, creates a number of psychological barriers which must be bridged if the teaching and learning process is to be a success. Moore and Kearsley (1996) explored transactional distance and considered it to be "the interplay between people who are teachers and learners, in environments that have the special characteristic of being separate from one another, and a consequent set of special teaching and learning behaviours" (200).

Therefore, when evaluating a discipline offered via online technologies, it is important that the above factors be adequately addressed.

Quality Evaluations at UWI

Campbell and Rozsnyai (2002) described evaluation as "any process leading to judgements and/or recommendations regarding the quality of a unit" (31). At UWI, the Quality Assurance Unit describes its evaluation process as "review of the quality assurance procedures that are in place, in a department, institute, school, centre, unit or site to determine whether the procedures are adequate to assure that the quality of provision and associated processes are maintained and enhanced" (UWI, Quality Assurance Unit 2010, 7). Although the Quality Assurance Unit's process is a developmental one, it does lead to a judgement about the quality of the entity and recommendations for improvement.

The university's quality evaluation process takes place every two to three years, usually mid-term and the year before a discipline review. The instrument used by the university is self-administered and consists of fourteen quality indicators. Each indicator is supported by several items of evidence that must be made available and connected to any judgements reached. The quality indicators are:

1. designation
2. mission/aims/objectives
3. organizational structure
4. course/programme development and review
5. staff meetings
6. student feedback and data
7. external opinion
8. student profile
9. learning objectives
10. student support
11. resources to support student learning
12. staff and staff development
13. quality assurance and enhancement
14. research, outreach and publication

The respondents can indicate whether the specific indicator is applicable (the instrument calls this "available"), provide comments or reports on action to be taken, indicate the person responsible for taking the action and identify an implementation date for the specific action. The completed instrument is submitted to the Quality Assurance Unit where it is reviewed for completion, evidence of quality and areas requiring further attention. Following this desk review, a meeting is scheduled between the unit and the entity being evaluated. During this meeting, the entity is required to provide additional evidence to support their self-evaluation claims. Additionally, there are discussions relating to the areas requiring improvement and possible solutions are investigated.

Although the current university evaluation instrument adequately addresses the quality indicators for the face-to-face programmes, there are two major quality indicators that are vital for online education which are not present, namely student engagement and interaction and technological support. Student engagement and interaction needs, as posited by Moore (1972) in his theory of transactional distance, are key contributors to student satisfaction and retention in online programmes. Additionally, in an online environment there is need to ensure that the department provides adequate technological support for both its faculty and students. An additional factor that may improve the effectiveness of the university's current evaluation instrument is the introduction of a weighting mechanism; this would allow departments to gauge performance holistically as well as in specific areas. This ability would allow departments to target resources to the specific areas requiring most quality improvement.

Shelton's Quality Scorecard

The quality scorecard for the administration of online education programmes is a tool that can be used to evaluate the quality of online programmes and its effectiveness is not affected by the size of the institution being evaluated. The scorecard is weighted. This allows the institution to gauge its performance based on the following scores:

- 90–99 per cent (189–209) – exemplary, little improvement needed
- 80–89 per cent (168–188) – acceptable, some improvement recommended
- 70–79 per cent (147–167) – marginal, significant improvement needed in multiple areas
- 60–69 per cent (126–146) – inadequate, many areas of improvement needed
- 59 per cent and below (125 and below) – unacceptable

The scorecard uses a four-point Likert Scale, where respondents are asked to rate the various statements as detailed in table 11.1.

Table 11.1. Shelton's (2010) rating statements and weighting score

Rating statement	Weighting score
Not observed	0
Insufficiently observed	1
Moderate use	2
Meets criteria completely	3

The scorecard provides the following explanation for the rating statement descriptors (Shelton 2010, 531):

- Not observed – the administrator does not observe any indicators of the quality standard in place.
- Insufficiently observed – the administrator has found a slight existence of the quality standard in place. Much improvement is still needed in this area.
- Moderate use – the administrator has found there to be moderate use of the quality standard. Some improvement is still needed in this area.
- Meets criteria completely – the administrator has found that the quality standard is being fully implemented and there is no need for improvement in this area.

The scorecard has seventy questions which are divided among the following categories:

1. institutional support
2. technology support
3. course development and instructional design
4. course structure
5. teaching and learning
6. social and student engagement
7. faculty support
8. student support
9. evaluation and assessment

The scorecard is a self-administered instrument for administrators of online programmes.

Although the quality scorecard is an effective instrument for the evaluation of online programmes, it does not currently allow the users to score the nine categories individually. Nonetheless, it is a tool that could be adapted by the UWI for the evaluation of its online programmes.

Comparison of the UWI Quality Evaluation Process with Shelton's Quality Scorecard

When the themes from the UWI and the quality scorecard were compared, it was found that the university's evaluation process was wider than the quality scorecard, at least as it related to themes. However, what was not clear was whether some of the themes and descriptors in the UWI's evaluation instrument actually had a significant impact on the quality or lack thereof of the programme and/or discipline. For example, the UWI theme "designation" is described as "to confirm categorisation (where relevant) of unit in accordance with the UWI definition of particular entity, department, institute, school, centre, unit or site . . . indicate (using the frame of reference in the UWI policy), the correct status of the entity being evaluated" (UWI, Quality Assurance Unit 2010, 1). However, there is no clear link between this theme, its components and the quality of the programme or discipline. This theme would not be relevant in an online programme evaluation instrument. Similarly, although the use of external opinion, student profile information and staff meeting data is useful for a department in revising or updating a programme, it is the outcome of the use of the data generated and not the data itself that is important. Therefore, these three domains of information would be better used as evidence to quality initiatives as opposed to actually being quality initiatives. Hence, they would not be applicable as quality indicators for an online evaluation instrument.

Differences between the UWI quality evaluation instrument and the Shelton scorecard are captured in table 11.2.

Table 11.2. Comparison between the UWI's evaluation and Shelton's quality scorecard themes

UWI evaluation themes	Shelton's quality scorecard themes
Designation	
Mission/aims/objectives	
Organizational structure	Institutional support
Course/programme development and review	Course development and instructional design
	Course structure
Staff meetings	
Student feedback and data	Evaluation and assessment
External opinion	
Student profile	
Learning objectives	Teaching and learning
Student support	Student support
Resources to support student learning	
Staff and staff development	Faculty support
Quality assurance and enhancement	
Research, outreach and publications	
	Social and student engagement
	Technology support

Table 11.2 shows that the UWI evaluation instrument and Shelton's quality scorecard have seven themes in common. However, as noted previously, it also shows that the UWI's evaluation instrument does not cover two key themes for online education, social and student engagement and technology support. First, social and student engagement, as previously noted in the discussion on transactional distance, is directly related to the psychological distance that online students experience and is a quality indicator that should be present in any evaluation of an online programme or discipline. Second, technology support is a key to the students' ability to effectively interface with the online technology. Volery and Lord (2000), when referring to the works of Sanders and Nagelhout (1995), indicated that "the reliability, quality and medium richness [of the technology] are key technological aspects to be considered" (218). Hence, if the technology aspect is that important to online learning then technological support must be equally important to the success and quality of the online offering. Therefore, for the evaluation of an online discipline or programme it is crucial that technological support is considered and analysed.

Conclusion

The quality assurance evaluation instrument is a key component in the assessment of quality of a programme or discipline. This instrument must be carefully designed to take account of the nature of the programme being evaluated. In the case of an online programme, the online learner has different needs from the face-to-face learner. The planning, development and evaluation of programmes should therefore reflect those varying needs. It is clearly more appropriate for an online institution to use a quality evaluation instrument that was developed from an online teaching and learning perspective. Presently, the UWI employs one general quality evaluation instrument. When this instrument is compared to Shelton's quality scorecard, which was developed specifically for online programmes, it was evident that the UWI instrument had weaknesses. Generally speaking many of the university's evaluation quality indicators are relevant for an online programme but it is vital that social and student interaction and technological support are also evaluated, given their key role in a quality online learning experience. Therefore, it is recommended that Shelton's quality scorecard be adopted and adapted as a regional quality evaluation instrument for online programmes/disciplines in the Caribbean. To achieve this, the following changes are proposed to Shelton's quality scorecard:

1. Rewording and/or deletion of some of the quality indicator statements to reflect Caribbean realities and language usage and understanding, for example, "policy and process" to replace "support ADA [Americans with Disabilities] requirements".

2. The weighting and categorizing of the scorecard to ensure that each theme can be scored separately. This would enable the various sections that may be responsible for discrete components to gauge its performance. In addition, it would assist with the easy detection of specific areas requiring improvement.

References

Benson, A.D. 2003. "Dimensions of Quality in Online Degree Programs". *American Journal of Distance Education* 17 (3): 145–49.

Campbell, C., and C. Rozsnyai. 2002. "Quality Assurance and the Development of Course Programmes". Papers on Higher Education Regional University Network on Governance and Management of Higher Education in South East Europe Bucharest, UNESCO.

Cassey, D.M. 2008. "A Journey to Legitimacy: The Historical Development of Distance Education through Technology". *TechTrends: Linking Research and Practice to Improve Learning* 52 (2): 45–51.

Daniel, J., A. Kanwar and S. Uvalic-Trumbic. 2009. "Breaking Higher Education's Iron Triangle: Access, Cost, and Quality". *Change: The Magazine of Higher Education*. http://www.changemag.org/Archives/Back%20Issues/March-April%202009/full-iron-triangle.html.

Dilbeck, J. 2008. "Perceptions of Academic Administrators Towards Quality Indicators in Internet Based Distance Education". PhD diss., Indiana State University.

Freire, P. 1970. *Pedagogy of the Oppressed*. New York: Continuum.

Harvey, L., and D. Green 1993. "Defining Quality". *Assessment and Evaluation in Higher Education* 18 (1): 9–34

IHEP (Institute for Higher Education Policy). 1998. *Assuring Quality in Distance Learning: A Preliminary Review*. Washington, DC: IHEP. http://www.ihep.org/assets/files/publications/a-f/AssuringQualityDistanceLearning.pdf.

———. 2000. *Quality on the Line: Benchmarks for Success in Internet-Based Distance Education*. Washington, DC: IHEP.

Lee, J., and C. Dziuban. 2002. "Using Quality Assurance Strategies for Online Programs". *Educational Technology Review* 10 (2): 69–78.

Linstone, H.A., and M. Turoff. 2002. "Introduction". In *The Delphi Method: Techniques and Applications*, edited by H.A. Linstone and M. Turoff, 3–12. Newark, NJ: New Jersey Institute of Technology.

Meyer, K.A. 2002. *Quality in Distance Education: Focus on On-Line Learning*. San Francisco: Jossey-Bass.

Mishra, S. 2006. *Quality Assurance in Higher Education: An Introduction*. Bangalore, India: NAAC.

Moore, M.G. 1972. "Learner Autonomy: The Second Dimension of Independent Learning". *Convergence* 5 (2): 76–88.

Moore, M.G., and G. Kearsley. 1996. *Distance Education: A Systems Review*. Belmont, CA: Wadsworth.

Schrum, L., and S. Hong. 2002. "From the Field: Characteristics of Successful Tertiary Online Students and Strategies of Experienced Online Educators". *Education and Information Technologies* 7 (1): 5–16.

Shelton, K. 2010. "A Quality Scorecard for the Administration of Online Education Programs: A Delphi Study". PhD diss., California State University-Hayward

University of the West Indies Strategic Plan, 2007–2012. Kingston: University of the West Indies.

UWI, Quality Assurance Unit. 2010. *Quality Assurance and Quality Evaluation*. UWI.

Vlăsceanu, L., Grünberg, L., and Pârlea, D., 2007, *Quality Assurance and Accreditation: A Glossary of Basic Terms and Definitions*. Bucharest: UNESCO-CEPES.

Volery, T., and D. Lord. 2000. "Critical Success Factors in Online Education". *International Journal of Educational Management* 14 (5): 216–23.

Yusuf-Khalil, Y. 2006. "Engendering Development Needs: 'Doing' Gender through Distance Learning in the English-Speaking Caribbean". Paper presented at the fourth Pan-Commonwealth Forum on Open Learning. http://pcf4.dec.uwi.edu/viewpaper.php?id=222.

12 | Massive(ly) Open Online Courses
Opportunity or Threat to Caribbean Higher Education?

PATRICK ANGLIN

> MOOCs, regardless of underlying ideology, are essentially a platform. Numerous opportunities exist for the development of an ecosystem for specialized functionality in the same way that Facebook, iTunes, and Twitter created an ecosystem for app innovation. I don't know if MOOCs will be transformative in higher education. I'm not sure that they'll be half as disruptive as some claim. They are, however, significant in that they are a large public experiment exploring the impact of the internet on education. (Siemens 2012, 2)

Massive(ly) open online courses (MOOCs) represent the latest innovation in the use of technology in distance education. According to Rodriguez (2012) and Daniel (2012), the term MOOC, for massive or massively open online course, was first coined by Dave Cormier and Bryan Alexander to describe an open online course offered, through the University of Manitoba, by George Siemens and Andrew Downes, in 2008. A MOOC, according to Daniel, "is a type of online course aimed at large-scale participation and open access via the web" (2012, 4). MOOCs have two main features that differentiate them from traditional distance or online education offerings:

1. MOOCs are geared towards a large number of participants – hence the term (M)assive.
2. They are available to a wide cross-section of individuals with the only criterion being access to the web by the participant – hence (O)pen and (O)nline.

This chapter introduces the MOOC, its evolution within the context of distance education and the use of information and communication technologies in education. The challenges and opportunities presented by MOOCs for higher education in the Caribbean are also explored.

The Evolution of MOOCs

MOOCs evolved out of the open educational resources (OER) movement, which began out of a programme spearheaded by the Hewlett Foundation in 2002. The goal of the Hewlett Foundation programme was "to use information technology to help equalize the distribution of high-quality knowledge and educational opportunities for individuals, faculty, and institutions within the United States and throughout the world" (Atkins, Brown and Hammond 2007, 2). This initial goal later evolved into the current open content initiative or the open educational resources initiative. Atkins, Brown, and Hammond define OER as "teaching, learning, and research resources that reside in the public domain or have been released under an intellectual property licence that permits their free use or re-purposing by others" (4). The term OER was officially generated at a UNESCO conference in 2002 and describes resources openly available to all persons and users do not need to pay royalties to use them (Butcher 2013).

Although MOOCs evolved out of the OER movement, a MOOC by definition does not mean open licensing. Several MOOCs do not implement an open licence concept and therefore do not subscribe to the definition of "open" as initially envisaged by the OER movement. While not initially described as such, Massachusetts Institute of Technology's OpenCourseWare Project was the first MOOC platform. OpenCourseWare, piloted in 2002 with 50 courses, currently houses materials from over 2,150 courses and has attracted more than 125 million visitors (OpenCourseWare 2013). Figure 12.1 shows the MOOC timeline to date.

The Evolution of Distance Education with That of Technology

The evolution of MOOCs cannot be isolated from the general evolution of distance education and its use of technology, specifically information and communication technologies, as enablers and supporting tools. Anderson and Dron (2011) cite the scholarship of Garrison (1985) and Nipper (1989), who organize the evolution of distance education into five generations. Each generation of the evolution corresponds to a particular technology and that technology influences the distance education practice.

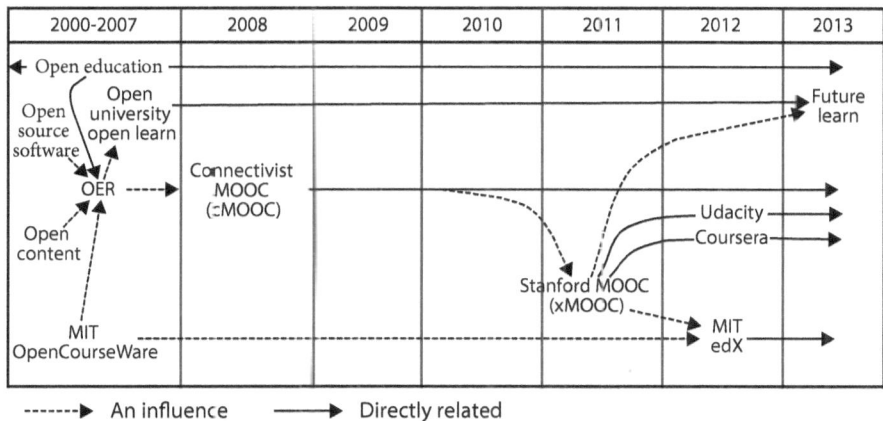

Figure 12.1. MOOC timeline
Source: Yuan and Powell (2013, 6).

Anderson and Dron clearly identify four generations, with the fifth being somewhat nebulous. The first generation of distance education utilized mail correspondence to deliver course materials to students. This was followed by the use of mass media such as radio and television, and film in the second generation. The third generation used interactive technologies such as audio, text, video, all followed by web and immersive conferencing. The fourth, and current, generation utilizes intelligent databases, Web 2.0 and semantic web technologies. Web 2.0 has been described as the use of dynamic technologies, as opposed to the static ones of the past, to enable the web to be a platform for delivering a greater set of services that allows greater interaction among users (Graham 2005; O'Reilly 2005; Ibm.com 2006). While semantic web technologies, also referred to as Web 3.0, are those that use basic artificial intelligence to allow devices, such as personal computers, to gather and interpret data provided by other web-connected devices, or websites, and then use these data to provide services to users, or devices, without input from a human user (Berners-Lee, Hendler and Lassila 2001; Shadbolt, Berners-Lee and Hall 2006).

In addition to distance education evolving with technology, the pedagogies also evolve in line with the technology. Anderson and Dron (2011) identify three generations of distance education pedagogies that match with the second, third and fourth generations of distance education technologies.

As outlined in figure 12.2, Anderson and Dron (2011) do not match the first generation of distance education, that is, the utilization of mail correspondence to deliver course materials to students, with any particular pedagogy and begin at the second generation. They posit that the first generation of pedagogy, cognitive-behaviourism, which focuses on the changes in learner behaviour in response to stimuli and led

Generation of distance education pedagogy	Technology	Learning activities	Learner granularity	Content granularity	Evaluation	Teacher role	Scalability
Cognitive-behaviourism	Mass media: Print, TV, radio, one-to-one communication	Read and watch	Individual	Fine: scripted and designed from the ground up	Recall	Content creator, sage on the stage	High
Constructivism	Conferencing (audio, video, and Web), many-to-many communication	Discuss, create and construct	Group	Medium: scaffolded and arranged, teacher-guided	Synthesize: essays	Discussion leader, guide on the side	Low
Connectivism	Web 2.0: Social networks, aggregation and recommender systems	Explore, connect, create and evaluate	Network	Coarse: mainly at object and person level, self-created	Artifact creation	Critical friend, co-traveler	Medium

Figure 12.2. Summary of distance education pedagogies

Source: Anderson and Dron (2011).

to interventions such as computer-assisted instruction and instructional system designs, matches the era of mass media (print, TV, radio) – the second generation of distance education. The next pedagogical generation, constructivism, developed in conjunction with two-way communication such as audio-conferencing, video-conferencing and web-conferencing. Social constructivist pedagogies, for example, emphasize the social nature of knowledge and that instructors do not merely transmit knowledge but that each learner constructs ways to create new knowledge and integrate this new knowledge with existing knowledge. The third generation of distance education pedagogy is connectivism, which emphasizes the building of networks of information, contacts and resources, all of which are used to find solutions to real problems. Connectivism assumes that learners are able to access networks and are literate and confident enough to use these effectively to assist their learning. Connectivism matches Web 2.0 technologies such as social media and their use in distance education.

Although MOOCs fall within the fourth generation of distance education technologies, there are basically two types of MOOCs: those which prioritize learning in informal networks, focusing on the learning process, and based on connectivism theory; and those that are content-based, and follow behaviourist approaches to learning (located in the second generation of distance education technologies). The learning process-based MOOCs are referred to as cMOOCs, for collectivist MOOCs, while the content-based MOOCs are referred to as xMOOCs or AI Stanford like courses (Liyanagunawardena, Adams and Williams 2013). Yuan and Powell (2013) state that xMOOCs place greater emphasis on connected collaborative learning while cMOOCs focus on the exploration of new pedagogies outside of those used in the traditional classroom. xMOOCs are more "traditional", focusing on extending the traditional pedagogical models of the institution to this new delivery mechanism. Institutions wishing to extend their brand to the MOOC market and commercial companies wishing to benefit from another income stream tend to set up xMOOCs, while the so-called educational purists tend to set up cMOOCs.

One of the easiest distinctions between xMOOCs and cMOOCs is that offered by degree of freedom, which states that xMOOCs are based on "professor-centric courses" normally associated with established universities and for-profit companies, whereby courses are delivered to enrolees by an instructor, while cMOOCs involve enrolees acting as both students and instructors. To further clarify, xMOOCs are normally used to deliver content to learners, while cMOOCs are used as tools in educational research to gather data. That said, the most clear distinction is provided by Siemens (2012), who states that cMOOCs focus on knowledge creation and generation whereas xMOOCs focus on "knowledge duplication". It is safe to say that most commercial MOOCs are xMOOCs.

Concerns with MOOCs

There are several concerns with current MOOCs. Referring to xMOOCs, Yuan and Powell (2013) raise concerns about the pedagogy and quality of current MOOCs. On the matter of pedagogy, Yuan and Powell question whether MOOCs follow pedagogical and organizational approaches to online learning for quality experiences and outcomes for students. Another pedagogical question concerns what new pedagogical approaches are necessary to maximize learning outcomes. In other words, should MOOCs use existing pedagogies or develop new ones? A further concern is quality assurance. Rodriguez (2012) makes the point that one of the greatest challenges for MOOCs is the assessment of what is being learned. He indicates that, even within the realm of open online courses, accreditation models are inappropriate. An important issue is the granting of credit when participants, some of whom are peripheral, may not be doing the same level of work that others do.

The Question of Quality

The question of quality in Caribbean higher education is not an inconsequential one. The Caribbean Area Network for Quality Assurance in Tertiary Education (CANQATE) is the regional organization established in 2004 to, among other matters, promote good quality assurance practices in the Caribbean region. CANQATE also shares information about the maintenance, evaluation, accreditation and improvement of higher education standards, and acts primarily through a network of external quality assurance agencies. According to the CANQATE website, nine territories have external quality assurance agencies recognized by the Caribbean Community. However, according to information obtained from the CANQATE Secretariat in October 2013, representatives from eighteen countries (seventeen Caribbean nations and the United States) participate in CANQATE conferences. In addition to information sharing, CANQATE, which is a sub-network of the International Network for Quality Assurance Agencies in Higher Education, provides training (professional development) in quality assurance to its members and also facilitates regional discourses on policy and research in quality assurance and quality enhancement.

Interestingly, the accreditation of online and distance content has barely begun. Although some Caribbean accreditation agencies are preparing distance education standards, few have implemented any such standards. In addition, while most, if not all, indigenous Caribbean universities offer online courses, few have completed the implementation of specific quality assurance mechanisms to govern these. There is a distinct gap in the quality assurance mechanism for online education. The literature focusing on the appropriateness of the pedagogical approaches taken by Caribbean institutions, as well as the appropriateness of the content used in distance education, is scant to non-existent.

Caribbean accreditation agencies should not only fast-track the implementation of distance education standards, on which the region is currently lagging, but now consider what impact the advent of MOOCs will have on these. The possibility of traditional distance education quality assurance standards being inadequate for MOOCs is a real one.

Siemens (2012), as seen in figure 12.3, highlights several other concerns grouped by whether these affect cMOOCs or xMOOCs. For cMOOCs, the concern is the appropriate revenue model to pursue in order to ensure financial viability. Although xMOOC business models are fairly straightforward, as discussed further in this chapter, the business model for cMOOCs, which focus on a different pedagogical approach, is less so. For xMOOCs, Siemens identifies three issues:

1. accreditation
2. course completion rate
3. student authentication

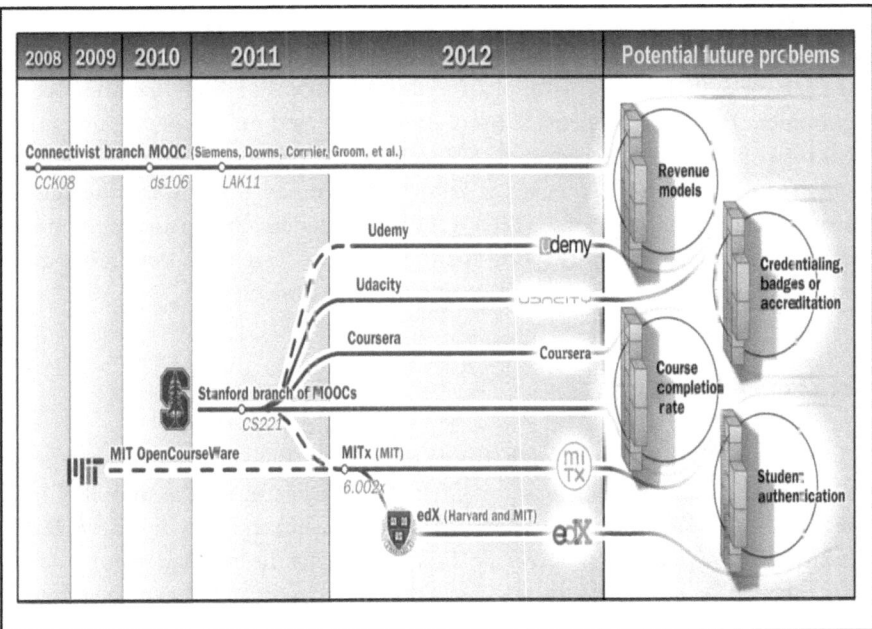

Figure 12.3. MOOC timeline and potential problems

Accreditation

The issues of accreditation raised by Siemens are similar to those raised by other contributors to the debate (Rodriguez 2012; Kim 2013) and have already been mentioned. In respect of course completion rates, it is acknowledged that while data is difficult to ascertain, MOOCs have a very high withdrawal/discontinuation rate.

Completion Rates

Liyanagunawardena, Adams and Williams (2013) quote statistics from Katyjordan.com (2013), who collated completion rates for twenty-four MOOCs studied, indicating that, as of 11 March 2013. "the highest completion rate achieved was 19.2%" (218). While the completion rates are low, it should be noted that enrolment in these courses tends to be in the tens to hundreds of thousands. Even with these extremely high rates of enrolment, it might be difficult for a Caribbean university to justify the use of scarce financial resources, provided, in some cases, by the people of the Caribbean, to offering a "free" course that most students fail to complete.

Identity

The issue of authentication is also a problematic one. For MOOCs, there is no requirement that students provide proof of identity. And unless there is a requirement for certification or recognition of credit, authentication is really unnecessary since the benefit to the learner is the knowledge accrued from taking the course. However, when, as in some of the MOOC business models, quality assurance mechanisms are required for entities external to the MOOC provider, determining the identity of tens of thousands of learners might prove difficult.

Insufficient Research

McAuley et al. (2010) sound another note of caution. They indicate that while the MOOC model has seemingly vast potential, it is nascent and is the subject of little research. Assessment, through research, of the current effects and a possible data-led forecast of future potential is necessary. This is especially the case in a region, such as the Caribbean, where, due to the scarcity of resources, in order to maximize resource use, expenditure must be based on options with proven track records. McAuley et al. (2010, 7) raise specific pedagogical issues, challenges and questions such as:

- the extent to which it (MOOCs) can support deep enquiry and the creation of sophisticated knowledge;
- the breadth versus the depth of participation;
- whether and under what conditions successful participation can extend beyond those with broadband access and sophisticated social networking skills;
- identifying the processes and practices that might encourage lurkers, or "legitimate peripheral participants", to take on more active and central roles;
- the impact or value of even peripheral participation, specifically the extent to which it might contribute to participation in the digital economy in extra-MOOC practices;
- specific strategies to maximize the effective contribution of facilitators in particular and more advanced participants in general.

MOOCs and the Future of Caribbean Universities

Globalization has affected many facets of Caribbean life, from the export of agricultural products to the influx of cheap, easily accessible, consumer products and food items. Higher education is no exception. According to Banchoff and Paul (2013), globally, more than 3.3 million students partake in higher education offerings outside their own country. The Caribbean does not factor in the statistics quoted by Banchoff and Paul, who indicate that the United States earns nearly $23 billion from higher education, while the countries, in addition to the United States, that attract the most international students are the United Kingdom, Canada, France, Germany and Australia. Although the statistics for the number, or proportion, of Caribbean students who study abroad is not readily available, one would be naïve to believe that, even if Caribbean students are not currently studying abroad in droves, they would not do so if studying at universities and colleges outside the Caribbean is made easier.

The advent of MOOCs has significantly increased the ease of studying at non-Caribbean universities. Even without being specific to the Caribbean, some say that MOOCs will severely alter the viability of traditional universities and that in the future, a critical mass of students will choose to study online. Harden (2013, 1) paints a grim picture of the future of universities, as currently structured, at least in the United States. He states: "In fifty years, if not sooner, half of the roughly 4,500 colleges and universities now operating in the United States will have ceased to exist . . . . The future looks like this: Access to college-level education will be free for everyone; the residential campus will become largely obsolete; tens of thousands of professors will lose their jobs; the bachelor's degree will become increasingly irrelevant."

If Harden is to be believed, MOOCs will have a game-changing effect on higher education in the United States. Further, extending Harden's logic, this revolutionary effect will also impact the Caribbean, as MOOCs are available online and accessible from anywhere and, in some cases, at any time. Thrift (2013) is, however, of a different, more tempered view. He posits that while information technology changes some things, it does not change everything. As a corollary, Thrift states that MOOCs will change some things, but not others. "Bricks and mortar are not going to disappear, but information technology will continue to advance the practice of higher education" (2). It is my own view that, while MOOCs will have a profound impact on higher education, especially the distance mode, it will not supplant the traditional "brick and mortar" university but will provide useful and effective supplementation to offerings to students.

Although centred in United States, other countries have risen to take the MOOC challenge. Countries such as Australia, Brazil, Britain, Canada, China, Germany, Mexico, the Netherlands, Singapore and Spain have all responded with their own MOOCs (Grossman 2013; Lewin 2013; moocs.com). Interestingly, Veduca, the first MOOC provider in Latin America, operating out of Brazil, still provides content sourced from American universities such as UC Berkeley, Harvard and Columbia, only providing Portuguese subtitles to these. To its credit, the University of Sao Paulo has partnered with Veduca to offer the first Latin American-based MOOC. Currently, no Caribbean university, or group of universities, offers MOOCs or has partnered with external MOOC platforms to offer courses as a part of a contributing group of universities. This means, therefore, that students, whether current or prospective, cannot take an online course offered by any native Caribbean university without paying fees.

Wagner (2013) identifies the adult learner as the preferred target market for MOOCs, indicating that their (MOOCs') convenience and flexibility are attractive to the working professional. Andrew Ng (co-founder of Coursera) agrees; he states, "MOOCs and traditional classrooms are different . . . One advantage of MOOCs is their convenience. Eighty per cent of students [who enrol in MOOC courses] already have a bachelor's degree. . . If you're a working professional and want to learn something new, it's a challenge to go to a class on Tuesdays and Thursdays . . . MOOCs give more access to education. You have much more control over your learning" (2). This provides both an opportunity for, and challenge to, Caribbean institutions. Currently, very few Caribbean tertiary institutions offer a fully online postgraduate degree option. The University of the West Indies, for example, up to September 2013, offered only seven graduate programmes via the online mode only. This therefore means that unmet need is mostly being filled by offshore institutions, with the possibility of offshore MOOCs also entering the market. The opportunity exists for Caribbean institutions to develop graduate-level courses and deliver these as MOOCs. Of course, the issues identified with this new model must first be tackled and overcome.

A Business Model

A possible reason for the absence of Caribbean higher education institutions from the MOOC space might be the lack of a clear business model that can be used to recover costs. Up to 2012, Young (in Daniel 2012) made the point that the largest MOOC, Coursera, out of the United States, with current membership of over four million subscribers, was yet to make a profit and was unsure of the business model to adopt to generate revenue. (Coursera subsequently turned a profit in April 2013.) The advice to institutions provided by Yuan and Powell (2013) can also be applied to Caribbean higher education entities. They recommend the establishment of a separate business unit to facilitate the development and subsequent commercialization of MOOCS.

Coursera, which began as a non-profit MOOC, has instituted a for-profit arm, which uses three innovative methods for generating revenue: a verified completion certificate; a signature track and an affiliate programme. Fain (2013) describes the fee-based path as targeted to students wishing to earn a completion certificate that may be used for purposes, such as obtaining employment, considered more tangible than proving they attended the course. Coursera partner universities, of which there are eighty-five, will, by the end of 2013, jointly issue certificates to students who pay for and complete select courses. The signature track verifies the identity of the student so that he or she can be awarded the certificate upon course completion. Coursera also receives up to 10 per cent of the purchase price of books, from Amazon.com, if its students make a purchase after a suggestion from a professor in a Coursera course.

Most Caribbean universities (and tertiary institutions) are still struggling "to keep the lights on" amid the current austere financial climate and may therefore balk at the idea of offering courses, which incur cost, without a commensurate revenue stream. However, MOOCs provide an opportunity to not only participate in a growing trend, but also to open their "doors" to the wider global community. In addition to that wider global community, the Caribbean diaspora is also a potential market and is a unique niche that might be difficult for non-Caribbean universities to enter. Of course, MOOCs will also benefit those in the region who are unable to afford the current fees charged by local universities. Although their foray in the MOOC arena might initially prove non-profitable, Daniel (2012) outlines eight potential business models listed by Coursera that might be useful. The business models would benefit the institution, whether it partners with a commercial MOOC platform provider, and enters into a revenue sharing agreement, or it sets up its own. Daniel's eight potential models are:

1. Certification – where students might pay for a certificate obtained at the end of a course.
2. Paid assessments – where students are allowed to enrol and take the course for free but would pay for examination invigilation and any course transcript indicating that they have completed a course. This is more rigorous than the first model and requires greater resources from the university or its partner to ensure that the learner is authenticated and is assessed under examination conditions similar to "normal" students attending the university.
3. Employee recruitment – this involves commercial arrangements between the university and industry where potential employers pay for student performance records instead of, or in conjunction with, the student paying. This may work best where specific skills are in high demand.
4. Applicant screening – similar to third point above, this model involves employers or other universities, who wish to shortlist applicants for jobs or placement in university programmes, approaching the university offering specific courses for a list of students matching specific criteria, for example, all students who received an "A" in Business Ethics.
5. Learning supplementation – this involves face-to-face tutoring (or private tutoring using web-based or other distance technologies) or the marking of assignments, which students would then pay for.
6. Commercialization of MOOC platform – if the university develops its own MOOC platform, it might decide to sell this to other universities or to industry for their use in course delivery or training programmes.

7. Sponsorships – where external parties, such as granting agencies, government, industry, and so on, pay for the cost of developing and delivering a course.
8. Tuition fees – this takes us back to the realm of traditional distance education; however, the only difference would be to scale of the offering, that is, the number of students expected to take the course.

The Caribbean Potential

Even with the several issues facing MOOCs generally, the Caribbean has some specific factors that make MOOC adoption attractive. Downes (2013) points to the fact that the rates at which students are accepted to study in the region's tertiary institutions are relatively low. This, according to Downes, signals that a significant proportion of the demand for tertiary education, provided by some Caribbean institutions, is unmet. MOOC adaptation, by Caribbean tertiary institutions, would provide an avenue for meeting this unmet demand. Although a significant proportion of unmet demand stems from inadequate capacity among the top institutions, much of it is due to inability to afford higher education on the part of potential students.

Johnstone (2013) states that the Caribbean region has a fundamental problem of balancing the availability of public funds with the revenue needs of publicly funded tertiary institutions. He goes further to articulate the conundrum, stating that, in order to expand participation, additional resources will have to be identified and applied to the tertiary sector to increase its capacity to meet this expansion. However, capacity building itself requires resources which, in light of shrinking or static government funding, have to be funded by increasing fees to students. Increasing fees to students who already cannot afford the current fees is self-defeating.

Although there are ongoing concerns with quality assurance, MOOCs might be a part of the solution to balance the several variables of capacity, affordability and equity in the Caribbean region. Compared to face-to-face delivery, technology-enabled delivery – and MOOCs in particular – is a significantly cheaper and farther reaching option available to Caribbean tertiary institutions for increasing access to tertiary education.

Conclusion

The fact that MOOCs are impacting the face of global higher education is unquestionable. Whether MOOCs are a passing fad; whether their current structure will see radical reorganization, in response to the several issues affecting them, remains to be seen. However, one stark fact which must be confronted is that MOOCs are growing and are currently providing an alternative for those not willing, or able, to engage traditional higher education as well as those wishing to try something different. Universities around the world, except those in the Caribbean, are taking notice and are reacting by launching their own MOOC platforms or are collaborating with commercial platforms. Most have not yet fully determined how to generate revenue, but they are engaging the process. Caribbean universities are yet to do so.

For learners, participating in a MOOC only requires Internet access and, according to data from the International Telecommunications Union, the percentage of individuals using the Internet averaged roughly 52 per cent, across CARIFORUM in 2012. The average ranged from a low of around 11 per cent in Haiti to a high of around 84 per cent for Antigua and Barbuda. It is therefore not unreasonable to infer that most of the region's population has the ability to access MOOCs. While additional research will need to be conducted to determine relative numbers, it is not far-fetched to infer that MOOCs are providing an alternative for some persons in the region. Therefore, if only to retain Caribbean learners, Caribbean universities must begin to enter the MOOC market. This might first be done by collaborating with current MOOC platforms offered by existing players, where the only cost incurred will be the cost to develop courses.

From the institution side, issues of copyright and ownership have to be resolved. The question of whether the institution or the academic owns course content created for the institution must be settled. Universities that partner with MOCC platforms retain copyright of course content and therefore the platform provider cannot claim any of the university's intellectual property. However, the interaction between the university and the academic in relation to copyright might be less straightforward. Many Caribbean academics, similar to academics in other regions, are yet to relinquish or even compromise rights to their intellectual property – even course materials. This is an issue which must be addressed and a proper solution found.

MOOCs seem to be a two-edged sword. They have the potential to at least market Caribbean universities to a larger proportion of the global population if these universities develop courses and offer them through either a university-owned MOOC platform or in collaboration with a commercial MOOC platform. However, they have the potential of luring Caribbean learners away from regional universities and have so far "reinforced existing hierarchies in higher education" (Kolowich 2013, 1). The universities of the Caribbean need to react and they must do so quickly.

Caribbean institutions can begin acting now, even if to test the MOOC market, with minimal financial investment. In early 2014, edX (founded by Harvard University), Massachusetts Institute of Technology and Google launched a freely available MOOC platform, mooc.org. The platform enables educational institutions, businesses, non-profits, instructors and course authors to create their own MOOCs. The platform provides several services, including course building tools, course hosting, the ability to deliver small private online courses, and a service to promote courses. This is an opportunity, at low cost, for Caribbean institutions to get on board. The region can no longer afford to wait and watch to see if the fad fades. The time for action is now – mooc.org is registering and all are welcome.

Appendix 12.1. MOOC resources

Category	Articles
Introductory	McAuley, Stewart, Siemens & Cormier 2010; Koutropoulos & Hogue 2012; Rodriguez 2012; Bremer 2012; deWaard 2011; Kop & Carroll 2012; Masters 2011; Roberts 2012; Mahraj 2012; Daniel 2012; Hyman, 2012
Concept	Martin 2012; Bull 2012; Kirkwood 2010; Ardis & Henderson 2012; Bonino 2012; Vardi 2012; Mehaffy 2012; Anderson & McGreal 2012; Butin 2012; Mahraj 2012; Mehlenbacher 2012; Anderson & McGreal 2012; Hyman, 2012
Case studies	Bell 2010a; Bell 2010b; Bremer 2012; deWaard, Abajian, Gallagher, Hogue, Keskin, Koutropoulos & Rodriguez 2011; deWaard, Koutropoulos, Keskin, Abajian, Hogue, Rodriguez & Gallagher 2011; Downes 2008; Fini 2009; Fournier, Kop & Sitlia 2011; Kop & Fournier 2010; Kop 2011; Kop, Fournier & Mak 2011; Kop & Carroll 2012; Koutropoulos, Gallagher, Abajian, deWaard, Hogue, Keskin & Rodriguez 2012; Levy 2011; Mackness, Mak & Williams 2010; Mak, Williams & Mackness 2010; Roberts 2012; Rodriguez 2012; Stewart 2010; Schrire & Levy 2012; Vihavainen, Luukkainen & Kurhila 2012
Educational theory	Bell 2010a; Bell 2010b; Butin 2012; Cabiria 2012; deWaard, Abajian, Gallagher, Hogue, Keskin, Koutropoulos & Rodriguez 2011; deWaard, Koutropoulos, Keskin, Abajian, Hogue, Rodriguez & Gallagher 2011; Downes 2008; Kop & Fournier 2010; Kop, Fournier & Mak 2011; Mackness, Mak & Williams 2010; Mak, Williams & Mackness 2010; McAuley, Stewart, Siemens & Cormier 2010; Rodriguez 2012; Stewart 2010; Tschofen & Mackness 2012
Technology	Anderson & McGreal 2012; Fini 2009; Kop, Fournier & Mak 2011; Kop & Carroll 2012; Mak, Williams & Mackness 2010; McAuley, Stewart, Siemens & Cormier 2010; Rodriguez 2012; Vihavainen, Luukkainen & Kurhila 2012
Participant focused	Chamberlin & Parish 2011; Kop & Fournier 2010; Kop 2011; Kop & Carroll 2012; Koutropoulos, Gallagher, Abajian, deWaard, Hogue, Keskin & Rodriguez 2012; Levy 2011; Mackness, Mak & Williams 2010; Mak, Williams & Mackness 2010; Stewart 2010
Provider focused	MacIsaac 2012; Mahraj 2012; Sadigh, Seshia & Gupta 2012
Others	Esposito (2012); Frank (2012)

MOOC Articles 1 (*Source*: Liyanagunawardena, Adams and Williams 2013); *Directory of MOOCs* (http://www.moocs.co/)

References

Anderson, Terry, and Jon Dron. 2011. "Three Generations of Distance Education Pedagogy". *International Review of Research in Open and Distance Learning* 12 (3): 80–97. http://www.irrodl.org/index.php/irrodl/article/view/890/1663.

Atkins, Daniel E., John Seely Brown and Allen L. Hammond. 2007. "A Review of the Open Educational Resources (OER) Movement: Achievements, Challenges, and New Opportunities". *Report to the William and Flora Hewlett Foundation*. http://www.hewlett.org/uploads/files/ReviewoftheOERMovement.pdf.

Banchoff, T., and E. Paul. 2013. *The Globalization of Higher Education*. https://gle.georgetown.edu/wp-content/uploads/2013/08/Globalization-of-Higher-Education-vers.-May-22.pdf.

Berners-Lee, Tim, James Hendler and Ora Lassila. 2001. "The Semantic Web". *Scientific American* 284 (5) 34–43.

Butcher, Neil. 2013. "OERs and MOOCs: Old Wine in New Skins?" *eLearning Africa News Portal: Perspectives on ICT and Education in Africa*. http://www.elearning-africa.com/eLA_Newsporta/oers-and-moocs-old-wine-in-new-skins/.

Daniel, John. 2012. "Making Sense of MOOCs: Musings in a Maze of Myth, Paradox and Possibility". *Journal of Interactive Media in Education* 3 (December): 18. http://jime.open.ac.uk/jime/article/view/2012-18.

Downes, Andrew. 2013. "Financing Tertiary Education in the Caribbean: The Case of the University of the West Indies". Presentation at the Caribbean Development Bank, Annual Board of Governors' Meeting, St Lucia, May 22, 2013. http://www.caribank.org/uploads/2013/05/The-Case-Study-of-UWI.pdf.

Fain, Paul. 2013. "Paying for Proof". *Inside Higher Ed*. http://www.insidehighered.com/news/2013/01/09/courseras-fee-based-course-option.

Garrison, D.R. 1985. "Three Generations of Technological Innovations in Distance Education". *Distance Education* 6 (2): 235–41.

Graham, Paul. 2005. "Web 2.0". Last modified November 2005. http://www.paulgraham.com/web20.html.

Grossman, Sara. 2013 "American MOOC Providers Face International Competition". *Chronicle of Higher Education: The Wired Campus Blog*. http://chronicle.com/blogs/wiredcampus/american-mooc-providers-face-international-competition/44637.

Harden, Nathan. 2013. "The End of the University as We Know It". *American Interest*, January–February. http://the-american-interest.com/article.cfm?piece=1352.

Ibm.com. 2006. Developer Works Interviews: Tim Berners-Lee. http://www.ibm.com/developerworks/podcast/dwi/cm-int082206txt.html.

Johnstone, D. Bruce. 2013. "Financing Tertiary Education in the Caribbean: The Elusive Quest for Quality, Capacity, Affordability, and Equity". Paper presented at the Fourteenth William G. Demas Memorial Lecture sponsored by the Caribbean Development Bank, Castries, St Lucia, May 21. http://www.caribank.org/uploads/2013/04/14-William-Demas-Lecture-on-Education.pdf.

Katyjordan.com. 2013. *MOOC Completion Rates*. http://www.katyjordan.com/MOOCproject.html.

Kim, S. 2013. "Employee Caught Outsourcing His Job". *ABC News*. http://abcnews.go.com/story?id=18230346.

Kolowich, Steve. 2013. "Google and edX Create a MOOC Site for the Rest of Us". *Chronicle of Higher Education: The Wired Campus Blog*. http://chronicle.com/blogs/wiredcampus/google-and-edx-create-a-mooc-site-for-the-rest-of-us/46413.

Lewin, Tamar. 2013. "Universities Abroad Join Partnerships on the Web". *New York Times* (online). http://www.nytimes.com/2013/02/21/education/universities-abroad-join-mooc-course-projects.html.

Liyanagunawardena, Tharindu, Andrew Alexandar Adams and Shirley Ann Williams. 2013. "MOOCs: A Systematic Study of the Published Literature 2008–2012". *International Review of Research in Open and Distance Learning* 14 (3): 202–27. http://www.irrodl.org/index.php/irrodl/article/view/1455/2531.

McAuley, Alexander, Bonnie Stewart, George Siemens and Dave Cormier. 2010. "The MOOC Model for Digital Practice". *Paper – Massive Open Online Courses – Digital Ways of Knowing and Learning*. https://oerknowledgecloud.org/sites/oerknowledgecloud.org/files/MOOC_Final_0.pdf.

Nipper, S. 1989. "Third Generation Distance Learning and Computer Conferencing". In Mindweave: *Communication, Computers and Distance Education*, edited by R. Mason and A. Kaye, 63–73. Oxford: Permagon.

OpenCourseWare, M. 2013. About OCW | MIT OpenCourseWare | Free Online Course Materials. http://ocw.mit.edu/about/.

O'Reilly, Tim. 2005. "What Is Web 2.0: Design Patterns and Business Models for the Next Generation of Software". *O'Reilly Media Inc.* http://oreilly.com/web2/archive/what-is-web-20.html.

Rodriguez, C. Osvaldo. 2012. "MOOCs and the AI-Stanford like Courses: Two Successful and Distinct Courses". *European Journal of Open, Distance and E-Learning*. http://www.eurodl.org/?article=516.

Shadbolt, N., T. Berners-Lee and W. Hall. 2006. "The Semantic Web Revisited". *IEEE Intelligent Systems* 21 (3): 96–101.

Siemens, George. 2012. "MOOCs Are Really a Platform". *eLearnspace*. http://www.elearnspace.org/blog/2012/07/25/moocs-are-really-a-platform/.

Thrift, N. 2013. *To MOOC or Not to MOOC – WorldWise – Blogs – The Chronicle of Higher Education*. http://chronicle.com/blogs/worldwise/to-mooc-or-not-to-mooc/31721.

Wagner, Vivian. 2013. "MOOCs: Tearing Down the Ivory Tower". *TechNewsWorld*. Accessed September 15, 2013. http://www.technewsworld.com/story/MOOCs-Tearing-Down-the-Ivory-Tower-78909.html#.

Yuan, Li, and Stephen Powell. 2013. "MOOCs and Open Education: Implications for Higher Education". *Cetis Publications White Paper 2013: WP01*. http://publications.cetis.ac.uk/2013/667.

13 | Building a Quality Institution of Higher Learning in a Small State
Issues, Considerations and Challenges

S. Joel Warrican

Throughout the Caribbean, there is a focus on tertiary education for citizens. In fact, in 1997, the Caribbean Community set a goal of an increase from 8 per cent to 15 per cent for the rate of participation in tertiary education among individuals in the 18–25 age group by 2005 (Peters and Best 2005). This date was later revised to 2015 after many of the countries in the region failed to meet this target (Smith 2011). For this reason, in many of the countries in the region, there has been a movement towards expanding and upgrading tertiary institutions. This movement includes the amalgamation of existing post-secondary institutions as well as the conversion of colleges to university status. Though several of the countries in the region met this target, others have had to make revisions based on their varying realities. For example, in Barbados, where this goal has been surpassed, they are striving to widen access even further, setting a new goal of one university graduate per household by 2020 (Arthur 2007). To this end, there is a proposal to amalgamate three of its tertiary institutions (the Barbados Community College, the Samuel Jackman Prescod Polytechnic and Erdiston Teachers' Training College) to form a university college (Peters and Best 2005). This amalgamation has not yet occurred, but the model of amalgamation has been pursued by other countries in the region. For example, all the independent countries of the Organisation of Eastern Caribbean States, with the exception of St Vincent and the Grenadines (SVG), had already amalgamated smaller independent post-secondary institutions to form national colleges. Thus, it was no surprise when in the early 2000s, SVG, in its quest to improve education provisions for its citizens, also went the route of amalgamation.

SVG is a multi-island small state in the Caribbean. This small state, like other such countries, has limited financial resources, but has shown a level of commitment to the development of its people through education. As is the case in most of the other small states in this region, SVG allocates between 17 per cent and 20 per cent of its annual budget to education, which translates to approximately 5 per cent to 7 per cent of its gross domestic product (Smith 2011). Though the proportion is sizable, the actual amount in dollars is relatively small. Thus, like other countries in the region, there is much to do with little funding. Indeed, Smith (2011) points out that this allocation is close to the threshold of what most governments in the region can afford, indicating that there is not much more that can be spent on education. Furthermore, he asserts that higher education is much more expensive than the other levels, and it is becoming increasingly so.

Despite limited resources, SVG government seems committed to providing educational opportunities for its citizens. Consequently in harmony with Caribbean Community's notion of expanded participation, the government set a target of one university graduate per household by 2025 (Gonsalves 2010). In order to achieve this goal, attention was turned to upgrading tertiary education in the country with the introduction of the SVG Community College Act in 2005 (Act No. 28 of 2005). Of significance is the fact that in 2005, universal access to secondary education was achieved in SVG. This is a significant precursor to attention to tertiary education as it makes more of the country's young citizens eligible to pursue education beyond secondary school. To meet the education needs of this expanding group, consideration is being given to widening access to education at the tertiary level and diversifying the programmes being offered.

Thus, in 2009, following the trend of other countries in the Organisation of Eastern Caribbean States and guided by the College Act, the SVG government turned to amalgamating and upgrading four existing post-secondary institutions (the SVG A-Level College, the SVG Technical College, the SVG Teachers' College and the SVG School of Nursing) to form the SVG Community College (SVGCC). Consequently, these institutions ceased to exist as independent entities and became the Division of Arts, Sciences and General Studies; Division of Technical and Vocational Education; Division of Teacher Education and Division of Nursing Education, respectively. SVGCC has the stated mission of fostering the holistic development of learners through the provision of tertiary education that enables them to contribute proactively to a changing society, function effectively in the workplace and pursue further studies (SVGCC 2009).

As can be imagined, this amalgamation, which became operational in 2009, was not a seamless process since each institution individually had its own ethos, identity, mission and way of doing things. Furthermore, public perception, based on certain historical factors that are addressed in the next section, added to the challenge. This chapter reports on some of the issues, considerations and challenges encountered on the journey (2009–13) to forging a single, unified institution out of four formerly independent ones; in so doing, the SVGCC may be viewed as a model for access to tertiary level education in a small state.

Potential Barriers to Home-Grown Tertiary Level Education

The countries of the Caribbean region generally have colonial pasts. Whether by the English, the French, the Dutch or the Portuguese, these countries became the homes and financial sources to European settlers, who brought their customs, values and systems with them. These they imposed on the colonized and, as colonizers do, managed to promote the idea that this imposed heritage was superior to home-grown varieties (Howe 2005). This legacy of colonialism is so strong that even though political domination no longer exists, some of the values and systems remain entrenched (Lewis and Simmons 2010).

From its early colonial history, opportunities for higher education in the anglophone Caribbean were sparse and usually available in the United Kingdom. Consequently, individuals who were recipients of higher education, and, as a result were held in high esteem in the Caribbean Community, were those whose outlook was influenced by their British education. Anything else was considered to be inferior. This perspective continued and its influence can be seen today, even though it has softened over the years with the advent of the University of the West Indies (UWI), which was established as a regional institution of higher education in 1948. Its genesis was underpinned by regional developments such as adult suffrage, representative government and political independence in the early 1960s (Miller 2007). The university provided opportunities for more people in the region to obtain education at the tertiary level, and many took advantage of this provision at that time. Later provision of higher education was augmented with the introduction of state-run institutions.

But the benefits of providing higher education at home are threatened by perceptions woven through the legacy of a colonial past. The fact is that to many Caribbean citizens, higher education acquired abroad is still prized above the home grown. Thus, home-grown institutions of higher education, especially those at the national level, must face the scrutiny, and sometimes harsh and unwarranted criticism, of those whose colonial perceptions lead them to devalue and distrust these institutions. This was a major barrier to break down in SVG when the four independent entities were forged into a home-grown community college.

In the face of this challenge, setting up and establishing a tertiary level institution that would earn the support and trust of the population called for transparency of actions, assurance of the quality of the programmes, as well as attention to issues that would be of interest to potential clients. Though there were several issues, this chapter deals with those of access, equity, relevance, quality and financing. These issues were selected because they were identified by Smith (2011) as having most impact on higher education in the region.

In all that was done, issues related to the change also presented barriers, and remained central to the amalgamation process. Fullan (2007) argues that for change to be successful there needs to be a shared understanding of the change among all who are to be affected. The need for such an understanding was very evident in January 2009 when the four independent institutions were merged, becoming the four divisions of the SVGCC. Though called a community college, the four divisions had the tendency to continue operating independently, holding on to the characteristics that made them unique. This is a common reaction to change where there is uncertainty and absence of a clear and shared vision: those affected by the change cling to the familiar (Marris 1986). This state of affairs interfered with the four previously independent institutions functioning as divisions within a unified institution. It was in this environment that, in 2009, I was invited to the post of director to work with the college to facilitate the process of full amalgamation. As an external agent of change, I was charged with breaking down the barriers and establishing the college as a tertiary institution that could address the current and future higher education needs of the country. This chapter describes some of my experiences and lessons learned over the four years of that appointment.

Facilitating Adjustment to Imminent Change

Paying attention to the change as a central concern in the amalgamation process was indeed very important to the forging of the independent institutions into the four component divisions of the college. Thus, in harmony with Fullan (2007), one of the first actions to be taken was the establishing of a shared vision within and across the divisions. Consequently, not only were consultations with the staff held within the individual divisions, but also across divisions. Staff from the different divisions sat together to discuss issues related to them as members of the unified institution. Concerns, shared and different, were heard by all and common solutions sought. This helped to remove the "them and us" perspective that kept the divisions separate from each other. Another purpose for using the consultative approach was to quell fears of job loss and help the staff to develop a sense of job security. One of the reasons some staff members resisted the change was the sense of insecurity and worry about reclassification of positions and the implication of these for salary, promotion and pensions.

Discussion of the terms and conditions led to clarification of these issues and the resolution of misinformation and misperceptions held by some of the staff. It also helped to reinforce that they were valued by the college and that their participation, viewpoints and inputs were important for the smooth transition to unity in the college. Finally the iterative, consultative process provided a forum for sharing information with staff members. For example, the vision for the college was shared and discussed. Staff members were afforded the opportunity to ask questions about the plans for the college and to raise concerns about their role in those plans; and the board and I had the opportunity to talk about the way forward for the institution. Indeed, even if staff members did not agree with all of the plans, at least, they had information.

Participatory Leadership

It became evident early in the process that tough decisions would have to be made and that the type of leadership applied would be crucial to success. Consequently, I decided to implement a participatory leadership style. To this end, a leadership team was created. This team included staff members from all the divisions, and from different levels. Individuals in pivotal positions, such as deans of divisions and heads of departments, were inducted into the team and the vision for the college sold to them, showing them how their divisions and departments would benefit from pursuing this vision. These team members then acted as liaisons with the general staff, making it possible for many perspectives to be heard and taken into account when decisions had to be made, thus ensuring that staff understood that they were important to the process. Indeed, the leadership team was not averse to adjusting decisions to reflect suggestions coming from staff. This made it easier for staff to accept other decisions that had to be made by top management solely. In addition, when these decisions had to be made, information was shared with the staff and all efforts were made to acknowledge their concerns, even when the decision was not reversed.

Consultation

Throughout the process, sharing of information and consultation were prominent and consistent. For example, when documents such as the strategic plan, the operational plan, a lecturers' handbook and a students' handbook were created, these all went through the iterative and consultative process before final versions were made available. These activities collectively helped with the unification process and added clarity to the direction in which the college was heading. This was also vital for garnering the support of the staff for the various initiatives that were to be implemented to ensure that this unified institution of higher learning provides the best possible opportunities for the citizens of the country.

In order to ascertain the opportunities that the citizenry wanted, consultation was also done. Information was gathered via both formal and informal means. The informal means included listening to people who stopped me on the street; listening to the voices of people on radio call-in programmes; and reading comments in local newspapers. But formal investigations were also carried out. For example, arrangements were made for visits to various communities, especially those in difficult-to-access areas as well as the sister islands, to find out from people how they thought the college could serve them. Both employed and unemployed individuals were consulted, since these groups tend to have different needs.

In addition, businesses and local employers were also consulted. The Division of Technical and Vocational Education invited a sample of these to a symposium in which they not only talked about what the college could do to serve their needs, but also what they could do to help the college to discharge its mandate of providing higher education in the country. Partnerships were suggested and explored. For example, discussion about internships for students was held and greater potential for this explored. Consultation was held with the regional UWI as well as with the University of Technology, Jamaica, to determine what was needed so that the programmes offered by the college could articulate into their full degree programmes. Furthermore, a recent labour survey (Hamilton 2010) was also consulted to determine national needs.

Out of the consultation with these stakeholders, major areas of change at the fledgling college emerged but, as stated earlier, this chapter only highlights issues relating to access, equity, relevance, quality and financing. In the next section, attention is turned to the activities relating to these main areas.

Addressing Issues Related to the Amalgamation

Access

One of the areas of concern raised by the public related to *access* to the college. Though the mandate of the college is to serve the higher education needs of all the citizens of the country, only one sector of the country was being served primarily – new school leavers. Based on the practices of the previously independent institutions, the programmes that were offered as well as the fact that the operating hours were from 8:30 a.m. to 5:00 p.m. perpetuated this situation. Thus, to better serve the country, this had to change; access to the college had to be widened. Steps taken to accomplish this included in the initial phase: (1) lengthening the opening hours of the institution; (2) expanding the programme offerings; and later, (3) introducing the e-college, online access to the college. Widening access in these ways saw an increase in the college's enrolment from 1,484 in the 2008–9 academic year to 2,367 in 2012–13, an increase of almost 60 per cent.

A number of actions related to the three broad steps listed above had to be taken. For example, rather than closing from 4:30 p.m. to 5:00 p.m., the college now remains open until 9 p.m. This allows access to employed individuals who want to upgrade their qualifications or prepare for a job change. In addition, programme offerings were upgraded and diversified. For example, more subject areas were added and there was an increase in the levels of programmes. More associate's degrees were introduced and full bachelor's degrees were offered through franchised agreements with institutions such as UWI, the Jamaica Theological Seminary and University of Technology, Jamaica. This makes it possible for more citizens to pursue higher education at home, at a fraction of the cost that it would take for them to do so abroad. It is long recognized that for small developing states to make advancements, there is a need for skilled labour (Peters and Best 2005; Smith 2011) and thus, programmes that address the dearth of citizens with technical and artisan skills (Hamilton 2010) were also upgraded and, where necessary, introduced.

Resources

Another area affected by the widening of access is resources, and fortunately, not always in a negative way. For example, in order to increase programmes offered, the library had to be upgraded and thus as of 2012, upwards of EC$1M (US$370,370) has been spent on this. Further, an EC$30.5M (US$11.3M) building programme was undertaken to upgrade existing facilities and build new ones to accommodate students in the Division of Arts, Sciences and General Studies and the Division of Teacher Education. Thus, not only are more citizens able to access higher education, but efforts are being made to ensure that there are state-of-the-art facilities available to them.

In relation to widening access, there is still more to be done. For example, because of its terrain and owing to the fact that SVG is a multi-island state, there are some citizens who may find it extremely difficult, if not impossible to access higher education in this country. The college's physical facilities are all located on mainland St Vincent, and generally in the area of the main town. Transportation and other issues impose restrictions on some who may want to pursue a programme of study at the college. To address the needs of such individuals, the college is preparing to widen access through its e-college. With the ever-expanding availability of computer and Internet facilities, it will soon be possible for interested individuals in remote locations to have access to college programmes via this medium. Providing and valuing such access to higher education speaks to the issue of equity.

Equity

Prior to the establishment of the SVGCC, and perhaps coloured by the colonial past, as said earlier, certain types of higher education were prized above others. Thus, for example, students attending the former SVG Technical College were not afforded the same respect as those attending the SVG A-Level College. Thus, students who were very successful at the secondary level shied away from the technical areas. This lack of respect was also reflected in the distribution of resources. In order to ensure that this area of national need was adequately met by the college, these perceptions had to be addressed. This speaks to issues of equity among the four divisions of the college.

Several steps were taken to promote equity across the divisions of the college. For example, previously, students who completed studies in the technical field were primarily awarded local certificates, which were not recognized by the regional institution, the UWI, for matriculation into its programmes. Thus, this qualification was not valued in the same way as the certificates awarded by the overseas entities such as the University of Cambridge Examinations Council. To raise the status of the Division of Technical and Vocational Education, programmes were converted to associate degrees, which articulate into UWI bachelor's degree programmes. Furthermore, to address the issue where national scholarships were only awarded to students who completed academic programmes in Division of Arts, Sciences and General Studies, criteria for such awards are being revised so that students from any division, who meet the specifications, can be eligible for the prestigious national awards. Thus, it will soon be possible for students in the technical field, as well as the professional areas of nursing and education, to also benefit from this provision.

In addition, through an outreach programme, academically able secondary school students who are interested in the technical field are being introduced to many opportunities available at the college and made aware that there are as many prospects in the technical areas as they are in the so-called academic areas. The expectation is that as more academically able students pursue the technical areas, the perception that technical subjects are for those who are not "bright" will be removed. This would encourage young people who avoided the technical areas because of that stigma to enter the field and hence contribute to the expansion of the skilled technical and artisan pool needed to carry the country forward. Thus, by adopting a more equitable distribution of resources across the division, the college is helping to break down some of the prejudices and stigma associated with certain paths to higher education, thus ensuring that whatever route citizens take, they feel that they have achieved something of value.

Relevance

No matter how many new programmes are offered, there would be little interest from the population if they are not perceived to be relevant to them. Based on the finding of the labour market survey (Hamilton 2010) and feedback from employers who attended the previously mentioned Division of Technical and Vocational

Education symposium, a number of programmes deemed relevant to development were introduced. For example, anticipating growth in the tourist industry, programmes in the hospitality field were introduced. To support the technical thrust, an associate degree in technical and vocational education was introduced with strands in Built Environment and Home Economics. Furthermore, programmes in Business Studies were introduced for those who wanted to enhance their job performance or establish a platform for further study.

Relevance was also seen in the areas of education and technology. In education, an associate degree in early childhood education was introduced in harmony with the government's thrust in that area; bachelor's degrees in literacy and mathematics education were offered to address two areas of need in the country's education system. These are areas that are deemed key to the development of the country. Another step to ensure relevance was the shift from the British General Certificate of Education A-level qualification to the more culturally relevant Caribbean Advanced Proficiency Examination qualifications available through the Caribbean Examination Council. The Caribbean Advanced Proficiency Examination qualification had been available since 1998, but the A-level college retained for, the most part, the General Certificate of Education A-level examinations. However, recognizing that relevance of education is vital to the development of small states, the shift was made.

As was mentioned earlier, higher education is costly and can put pressure on the resources of small states. Consequently, the returns from this investment should be substantial enough to contribute to the development of the country as well as of the people as individuals. Thus, relevance of programmes is vital, and steps are being taken to ensure that, whatever programmes are offered by the college, relevance is a key factor. The SVGCC is committed to being a model for ensuring relevance of higher education in the Caribbean region.

Ensuring Quality

The perception that home-grown programmes are not of the quality of those available from elsewhere plagues national institutions of higher education in the Caribbean (Howe 2005). Thus, there was the need to raise awareness that, although the associate degrees awarded by the college are home-grown, they must meet certain standards that would allow students who pursue these qualifications to pursue bachelor's and postgraduate degrees elsewhere.

In order to ensure that the programmes offered by the college were of acceptable quality, consultation was done with regional and international universities that were potential partners. The aim of these initial consultations was to ascertain what they would be looking for in a community college associate's degree that could articulate into their own programmes. Based on the findings from these discussions, programmes were designed. Another step taken to ensure quality was the invitation to University of the West Indies and University of Technology, Jamaica, to do an assessment of programmes to determine how well they could articulate into their own bachelor's degree programmes. In addition, a quality assurance office was set up at the college to monitor programmes and a procedures manual was produced (SVGCC 2011).

Quality assurance for the college's programmes involves a number of activities. These include: (1) establishing an academic council through which all new courses must pass before they are introduced; (2) reviewing and revising existing programmes to make them current, relevant and up to par; (3) providing in-service training opportunities for staff so that they can upgrade their skills as instructors; (4) supporting leave for academic staff to pursue higher degrees in their field of interest; (5) upgrading equipment and facilities to state-of-the-art status; (6) carrying out departmental and divisional self-studies to identify areas of strength and areas of need and (7) applying to the local accreditation board for institutional accreditation, thus ensuring that an organization external to the college is also monitoring its programmes.

All of these activities were aimed at ensuring that the locally designed programmes meet international standards, while at the same time being relevant to the local context. As an added measure to the above-mentioned steps, a system of appraisal is being introduced to ensure accountability. This will help to ensure that the programmes on paper are being adequately translated into practice in the classrooms. These measures should serve to raise confidence within the population. They should feel that any programme that they decide to pursue with the college is not only worthwhile at home, but also abroad.

Financing

Finally, there is the issue of financing. Financing a national institution of higher learning is expensive, and as mentioned earlier, most of the Caribbean small states are at their threshold for financing education (Smith 2011). The challenge therefore is to provide the best possible services on a limited budget, without making costs burdensome for the clients. Consequently, in harmony with government policy, first time school leavers who are accepted into the college pay only minimal administrative fees, while other individuals who qualify for entry to any of the college's programmes pay tuition fees. However, these fees are substantially lower than what they would pay to pursue such studies abroad. The fees collected from students defray the recurrent costs of operating the institution, but financing for infrastructural development is largely provided directly by government or through donor agencies such as the European Union and the Caribbean Development Bank.

As mentioned previously, transparency and collaborative involvement of all the stakeholders are essential to building a quality institution of higher learning in a small state, especially when the actions taken require students, staff and the public to change their perceptions and behaviour towards the institution. The SVGCC is an example of how facing the challenges and seeking viable solutions can result in benefits for the institution itself and for those who use its services. As a result of its actions, the college is now operating a unified organization, responding to national and individual needs and working with other institutions to provide higher education of the best possible quality, given its financial limitations.

Several Sites, One College

One of the biggest challenges with establishing the community college out of four independent institutions was the fact that these institutions are situated on four different locations. As independent entities, each institution had its own traditions. For example, students wore uniforms that identified them with the particular institution with which they were registered; and graduation ceremonies were planned and executed individually. Despite the fact that these institutions were amalgamated to become divisions of the college, their being at different locations encouraged the tendency for them to continue to hold on to their individual traditions. Steps had to be taken to foster togetherness in the operations.

One of the first steps towards unification of the divisions was the formulation of common rules and regulations for all students, thus bringing all divisions under the same governance. To further strengthen the amalgamation, during the first year of my tenure, despite resistance from some staff members, the college had a single, much-publicized graduation ceremony, the start of a tradition as a single unit. Another activity designed to promote unity was the introduction of a common uniform for all divisions. A number of other initiatives were introduced to promote cohesion within and across divisions. For example, inter-divisional sporting competitions were introduced; regular faculty meetings are held, where staff members from the various divisions attend. The college also started hosting a lecture series where the public was invited to hear individuals from various professions and walks of life present on a topic of national interest. A particularly laudable aspect of the lecture series was that, for each lecture given, a student from one of the four divisions was selected to give a five-minute "mini lecture" related to the guest speaker's topic. This not only got the students involved in this college event, but it also showcased for the public the talent and potential across all of the divisions within the college. In the future, other unifying activities will be initiated.

For example, plans are underway to make it possible for students registered in one division to take courses in other divisions. With these unifying activities, the college is now being seen by all as a single, unified organization, even if it is spread across different sites.

Providing Quality Higher Education in a Small State: Lessons Learned

There is no doubt that past experiences shape present realities. This is very evident in the case of education, and specifically higher education, in SVG. The influence of two particular aspects was evident: the country's colonial past and its small state status. The SVGCC experience can be instructive to other countries with similar challenges. Some of the lessons learned are summarized next.

The Importance of a Shared Vision

Perhaps one of the most important lessons learned from the establishing of the SVGCC is the need for a shared vision among the leadership team. This vision should be one that stretches beyond what *is* to what *could* be; indeed, what *should* be. This means that there is a working towards a common goal that, though it may not always be clear to others, is so clear to the leaders that they have a good sense of what needs to be done to achieve it. This vision should be so real to the leadership team that they are able to sell it to others in order to garner their cooperation and commitment; they should be able to do so in such a convincing manner that even though those who are expected to implement the initiatives are hesitant, they are also excited by what they are hearing. It is also important that while sharing the vision for forward movement, attention should be paid to the concerns expressed by those who are to implement the various activities (Evans 2001; Fullan 2007). Furthermore, wherever possible without significantly diminishing that vision, alterations can be made to address these expressed concerns.

The Importance of Strong Leadership

There are some remnants of a colonial past that are deeply entrenched in the psyche of the formerly colonized and it takes bold, decisive actions to break through these barriers when innovations are introduced. One necessary feature therefore is a strong and committed leadership team, including key people from all levels. This team should share the vision for the institution and must be committed to the implementation of initiatives that will contribute to the success of the innovation, even if these initiatives are unpopular. However, small state phenomena such as political polarization and managed intimacy (Lowenthal 1987; Bray 2011) often make it difficult for leaders to make hard decisions. A major lesson learned is that for better results, leadership should involve as many individuals as is practical, even though in some cases, final decisions may rest with those at the top. This is most effective when there are clear procedures for making decisions that transcend individuals (Hargreaves and Fink 2006), and each member of the leadership team should understand who is responsible for making and enforcing decisions, not merely the person, but the position of the decision maker.

Quality Comes at a Price

Another valuable lesson learned during the higher education innovation relates to persuading staff members to make changes to their practices. Even when the need for the changes is understood, and individuals agree that actions must be taken if improvements are to be realized, they may not be willing to pay the price of making adjustments to their practices or beliefs. The lesson learned here is that commitment to quality higher education is no guarantee that those required to make changes will do so willingly, especially when it forces them out of their comfort zone. Thus, to achieve the goal of excellence in higher education, all involved must be urged to adopt their "new normal", even if the price is initial uncertainty and lack of confidence. Including individuals in pivotal positions in the leadership team who provide understanding support can allay anxiety but in some cases, firmness may also be required.

Costs and Value for Money

One of the prohibitive factors associated with higher education is cost. This is particularly true in small economies where personal and governmental funds are limited, especially in periods of harsh economic conditions such as have been evident in the early 2000s. Indeed, one of the challenges the college faced was to find ways to make higher education affordable to citizens in all economic brackets. Despite the low cost, however, the college had to convince the public that the services were worth the money spent; that is potential clients had to be reassured that the qualifications they would receive would have currency both at home and abroad. The lesson learned is that no matter how affordable higher education becomes, if it is not valued by the community for which it is intended, the desired impact will not be realized. In a society that values externally sourced, expensive education, homegrown higher education institutions that offer more affordable options should therefore be prepared to promote their product vigorously to the intended clients.

Entrepreneurship

Despite the fact that higher education is beneficial to small developing states, it can put a strain on public funds. Though governments in small states may provide financial support for such institutions, it is often not sufficient to support all that is needed to maintain high-quality services. Thus, national institutions of higher education in small developing states must find ways of bringing in additional income to supplement subventions from government. The lesson learned is that although institutions of higher learning in small states provide a vital community service, leadership may have to be bold and identify entrepreneurial pursuits that can supplement governmental funding. However, individual context must be considered and each institution must find a way of doing so in a sustainable manner that does not set costs outside of the reach of the intended clients.

Conclusion

The purpose of this chapter is to highlight issues, considerations and challenges associated with establishing an institution of higher learning in a small state, based on my experiences leading the home-grown community college in SVG. Issues relating to small state phenomena, colonial history and the change process are evident in the unfolding story of the SVGCC. In some cases these issues presented challenges as the college's administrators sought to widen access to this level of education, promote equity across divisions; ensure relevance and quality; secure finances and foster unity among divisions. In addressing the emerging issues, valuable lessons were learned. The SVGCC is still a work in progress, but after four years with this institution, I am convinced that it is on a path to make a sterling contribution to the development of this small multi-island state, and becoming a model for home-grown institutions of higher learning in the region.

References

Arthur, Owen. 2007. Keynote address. Opening plenary session of the Conference on the Caribbean. Washington, DC. June 19.

Bray, Mark. 2011. "The Small-States Paradigm and Its Evolution". In *Tertiary Education in Small States: Planning in the Context of Globalization*, edited by Michaela Martin and Mark Bray, 37–72. Paris: IILP/UNESCO. http://unesdoc.unesco.org/images/0021/002121/212196e.pdf.

Evans, Robert. 2001. *The Human Side of School Change: Reform, Resistance and the Real-Life Problems of Innovation*. San Francisco: Jossey-Bass.

Fullan, Michael. 2007. *The New Meaning of Educational Change*. 4th ed. New York: Teachers College Press.
Gonsalves, Ralph E. 2010. "Lifting the Education Revolution to the Next Level". http://ralphegonsalves.org/uploads/LIFTING_THE_EDUCATION_ REVOLUTION_ TO_THE _NEXT_LEVEL_1_.pdf.
Hamilton, T. 2010. *The Labour Market and Investment Study in St Vincent and the Grenadines*. (Report commissioned by Invest SVG.) Kingstown: Invest SVG.
Hargreaves, Andy, and Dean Fink. 2006. *Sustainable Leadership*. San Francisco: Jossey-Bass.
Howe, Glenford. 2005. *Contending with Change: Reviewing Tertiary Education in the English-Speaking Caribbean*. Caracas: International Institute for Higher Education in Latin America. http://unesdoc.unesco.org/images/ 0013/001315/131593e.pdf.
Lewis, Theodore, and Lynette Simmons. 2010. "Creating Research Culture in Caribbean Universities". *International Journal of Educational Development* 30: 337–44.
Lowenthal, D. 1987. "Social Features". In *Politics, Security and Development in Small States*, edited by Colin Clarke and Tony Payne, 26–49. London: Allen and Unwin.
Marris, Paul. 1986. *Loss and Change*. Rev. ed. London: Routledge and Kegan Paul.
Miller, Errol. 2007. "Research and Higher Education Policies for Transforming Societies: Perspectives from the Anglophone Caribbean". Paper presented at the regional seminar Research and Higher Education Policies for Transforming Societies: Perspectives from Latin American and the Caribbean, Trinidad and Tobago, July 19–20. UNESCO Forum on Higher Education, Research and Knowledge.
Peters, Bevis, and Gladstone A. Best. 2005. "Enhancing Access to Tertiary Education in Barbados: Promises to Keep". *Journal of Eastern Caribbean Studies* 30: 1–40.
Smith, Warren W. 2011. "The Paradigm Shift in Higher Education: A Call for Action". Speech delivered at the University of the West Indies, Cave Hill, July 7. http://www.caribank.org /uploads/publications-reports/statements-and-speeches/Speech_The_Paradigm_Shift _in_Education_ Final_Website_%20edited_26Jul2011.pdf.
SVGCC (SVG Community College). 2009. *Strategic Plan 2009-2014*. Kingstown: SVGCC.
———. 2011. "Quality Assurance Policies: A Procedural Manual" (draft). SVGCC.

14 | The University of the West Indies
Moving Quality to the Next Level

SANDRA GIFT

In 1994, the University of the West Indies (UWI) *Chancellor's Report on Governance* ushered in a strengthened system of quality assurance for UWI, focused on the institution's academic platform, and based on the "fitness-for-purpose" approach to quality. Since then, a heightened concern for quality and quality improvement has been at the forefront of the continuing growth and development of the university and, indeed, of higher education generally in the anglophone Caribbean. More than ever before, parents and students expect the best quality higher education student experience, both inside and outside the classroom. In addition, all key stakeholders expect UWI to continue to demonstrate its relevance to regional development. While the "fitness-for-purpose" concept of quality remains very relevant to quality assurance at UWI, given the imperatives of accountability and competitiveness, it is nonetheless appropriate and timely to consider quality standards and models that can contribute to UWI's institutional excellence in respect of both its academic and administrative platforms – moving quality to the next level. The literature on quality provides examples of quality standards and models, used in other sectors that may be considered for their potential to contribute to moving quality in a higher education institution to the next level. This chapter discusses the relevance of selected quality standards and models to the university.

Methodology

The methodology used in this chapter is the instrumental case study approach, within the qualitative study paradigm, which facilitates focusing on an issue for better understanding. The actual case is not so much the UWI as an institution as it is the issue of enhancing the institution's quality management system (QMS) so that it can better assure the quality of the learning experience of future students and, indeed, better serve all its stakeholders. The backdrop for the discussion is student feedback data for academic years 2011–12 and 2012–13 that point to institutional weaknesses that negatively impact the quality of learning but that also have implications for wider institutional effectiveness. Definitions of key concepts used in this discussion are presented below. (Definitions are drawn from several sources listed in the reference list).

Definitions

Quality Management System: A QMS is a set of interrelated or interacting elements that organizations use to direct and control how quality policies are implemented and quality objectives are achieved.

Standards: Standards are *statements* regarding an expected level of requirements and conditions against which quality is assessed or that must be attained by higher education institutions and their programmes in order for them to be accredited or certified. The term *standard* means both a fixed criterion (against which an outcome can be matched) and a level of attainment. Standards may take a quantitative form, being mostly the results of benchmarking, or they may be qualitative, indicating only specific targets (for example, educational effectiveness, sustainability, core commitments, and so on).

Model: A model is a simplified version of a system, the objective of which is to facilitate understanding by eliminating unnecessary components. A model contains only those features that are of primary importance to the model maker's purpose.

Student Feedback Data

In the 2011–12 and 2012–13 academic years, 986 students, across eleven programmes reviewed on the St Augustine campus, provided feedback on strengths and weaknesses of their programmes. The sampling methodology was a sample of convenience or non-probability sampling technique. Participants were asked to fill out an instrument that consisted of questions on their demographic characteristics (e.g., sex, level of study) and perceptions of the programme under review. In five open-ended items, they were asked to identify strengths, weaknesses, most valued aspects and most disliked aspects of the programme being reviewed and to list ways in which the programme could be improved. The majority of respondents were female and undergraduates. The programmes reviewed were Chemistry, Clinical Medical Sciences, Economics, Electrical and Computer Engineering, Geomatics and Land Information, History, Pharmacy, Psychology, School of Dentistry, School of Education and Sociology. The top three programme strengths respondents identified were syllabi and resource material (37.6 per cent), knowledge and competence of lecturers (18.5 per cent) and practical work/hands-on training (15.4 per cent). Students' open-ended comments identifying strengths included, in the area of syllabi and resource material: "courses are in-depth and applicable"; "my elearning has helped in doing review and keeping abreast [of] courses"; and "course content is beneficial for [the] future". In the area of knowledge and competence of lecturers: "excellent lecturers"; "very informative, well experienced lecturers"; and "lecturers are highly qualified". In respect of practical work/hands-on training: "the opportunity for hands-on learning"; "good exposure to a wide range of clinical learning and application"; and "clinical experience with a variety of procedures".

The main weaknesses respondents identified were certain aspects of syllabi and resource material (19.1 per cent), workload (13.6 per cent) and certain aspects of practical work/hands-on training (13.2 per cent). In the area of syllabi and resource material, in respect of particular programmes, weaknesses identified included the following: "too much emphasis paid to insignificant things"; "not enough is taught on an international level"; and "some courses are not detailed enough".

Where workload was concerned, weaknesses noted included: "workload extremely high for a semester course"; "heavy course loads that should be stretched to a longer period in order to facilitate learning"; and "too much information to assimilate at once". In relation to practical work/hands-on training particular weaknesses identified in some programmes included primarily issues relating to optimum functioning of laboratories. Generally, the qualitative findings, based on this student feedback, have implications for the organization and structure of programmes and the learning environment.

An Overview of Selected Quality Standards and Models

Addressing programme weaknesses identified by students requires a mix of both academic and administrative quality assurance processes and procedures that must be consistently defined and deployed if they are to be effective. A number of quality standards and quality models, no doubt, possess components that can be useful in fashioning an integrated QMS for a higher education institution. This chapter highlights, however, components of the International Organization for Standardization (ISO) 9000 series, which comprises several standards.

International Organization for Standardization (ISO) 9000 Series

The series includes ISO 9000, the most common standard for quality systems, and ISO 9001:2000, QMSs in education. ISO 9000 is the international standard for QMSs (developed for industry and reviewed by Shutler and Crawford (1998) for its application to higher education). These ISO standards together constitute the principal set of guidelines for a proposed eclectic quality model to move quality at UWI to the next level. With reference to ISO 9001:2000, ISO notes that its standard is not intended to have all educational institutions adopt a uniform approach to QMSs and recognizes that institutions may wish to go beyond ISO 9001:2000 requirements for greater efficiency of their QMSs. Given that ISO promotes using a process approach when developing, implementing and improving a QMS, the organization further proposes applying the "Plan–Do–Check–Act" (PDCA) quality model to all processes (ISO 2003). Inevitably, this suggests that ensuring the effectiveness of a QMS for particular institutions may require adopting components of more than one quality standard or model, thus allowing for an eclectic approach that could meet the specific needs of different kinds of educational institutions.

This implied flexibility is particularly valuable to a large, regional, multi-campus higher education institution such as UWI. The chapter therefore discusses the following models as being particularly useful given their simplicity and compatibility with existing features of the university's QMS: "Plan–Do–Check–Act", the Excellence in Higher Education Model, an adaptation of the Baldrige Framework and Root Cause Analysis. It is not possible here to engage in a detailed discussion of the characteristics of each of these standards and models. The discussion will focus on the ISO 9000 standard but will also consider key components of ISO 9001:2000 and of each of the selected models that can serve as inputs to an eclectic UWI quality model to achieve an integrated and enhanced QMS for the university.

Defining the Product of an Education or Training Institution

In the context of ISO 9001, the product of an education or training institution is "the enhancement of competence, knowledge, understanding or personal development of the student resulting from the learning experience" (BSI Quality Assurance 1995, cited in Shutler and Crawford 1998). The term "customer" is applied to any student or other stakeholder purchasing the institution's service. The process the student enters is the teaching delivered by academic staff according to the syllabus. Examinations are viewed as a quality control process "to verify that the students have actually learnt what the course was designed to teach them. Students who fail the examinations either leave the institution without the qualification or pass back into the teaching system to be re-examined" (Shutler and Crawford 1998, 155). The term "examinations" is used to refer to the various forms of assessment of student learning.

Requirements of ISO 9000

Shutler and Crawford (1998) review the requirements of ISO 9000, commonly used in industry, and how they may be applied in the higher education sector. Interpreting the language of industry in the context of higher education, they propose several requirements for a higher education institution based on ISO 9000. These requirements are: identify desired learning outcomes; assess students to ensure they satisfy appropriate entry requirements for the course; check the syllabus to ensure it results in learning outcomes desired by students; specify the teaching process; assign teaching duties based on training or experience and identify lecturers' training needs; ensure desired learning outcomes are achieved and give clearly defined responsibility and authority to administrators of the teaching system. These requirements, discussed below, directly impact the quality of teaching and learning.

This focus on consulting the learner stems from Shutler and Crawford's (1998) position that in fitting the industrial model to higher education, the focus must be placed on two key terms: "customer" and "product". Citing the guidance notes on the application of BS EN ISO 9001 for QMSs in Education and Training (BSI Quality Assurance 1995) they define the product of an education or training institution as "the enhancement of competence, knowledge, understanding or personal development of the student resulting from the learning experience". The customer is "any student, public organisation, or industrial body purchasing the service from the institution". Since the customer is the student, then the student is to be consulted in this model. There are likely to be other models that see the employer as the customer, for example, in the area of technical and vocational education and would thus give priority to consulting the employer. Ideally, both learners and employers should be consulted.

Identify Desired Learning Outcomes

ISO 9000 requires that higher education institutions identify the learning outcomes that students and other stakeholders want. While respondents to the quality assurance surveys mentioned above did not specify the learning outcomes they desired, their comments are reflective of the kind of learning opportunities they would like to have as well as the quality of the learning environment they desire – one that is well organized and structured, with good communication between teaching staff, administrative staff and students.

Students' recommendations to address areas of weakness in syllabi and resource material, organization and structure, and workload that could have implications for desired learning outcomes included the following: "more rotation"; "a clear syllabus should be given to students!"; "some parts of the syllabus could be excluded"; "greater knowledge base needed for community pharmacy": "greater focus on the clinical aspect"; "more therapeutic psychology"; "the university should provide a physics course in year 1"; and "lessening the course loads".

Without implying that everything students indicate they would like is necessarily valid, UWI would be moving quality to the next level by systematically and consistently engaging in consultation with students and potential employers regarding desired learning outcomes. This would also serve as testimony to students that they are truly at the centre of planning, teaching and learning. Acting as soon as possible to correct weaknesses identified by students would be in keeping with the requirements of this quality standard.

Assess Students to Ensure They Satisfy Appropriate Entry Requirements for the Course

The intent of this requirement is to ensure that students have the necessary knowledge and skills for successful completion of their selected programmes. In the case of the St Augustine campus, the perception is that UWI is in compliance with this requirement. Some measure of specific knowledge is required for entry into many UWI programmes. Consequently, only particular subjects are considered for specific programmes. There are minimum UWI matriculation requirements and, across faculties, academic ratings are consistently above these requirements (UWI 2010).

Yet, students' recommendations suggesting that this requirement might be an area in need of improvement included: "there should not be such a disparity in the competence level among students"; "[have an] admissions interview"; "require prior pharmacy work experience"; and "entry requirements may have to be revised to include A-level Comp Sci and Applied Math". The fact that some students assess their colleagues' competencies as below par suggests the possibility of a gap in quality to be closed in some programmes.

Check the Syllabus to Ensure It Results in Learning Outcomes Desired by Students

Whether by testing a prototype of the course with a small group of students, engaging in analysis of learning outcomes or undertaking comparisons with best practice courses, Shutler and Crawford (1998) specify the need to check syllabi to ensure they will result in learning outcomes desired by students. Implementation of this specification will serve to take UWI to the next level of quality.

Some students' recommendations to address areas of weakness in syllabi, resource material organization and structure, and workload that have implications for ensuring that syllabi result in learning outcomes desired by students are presented under the requirement "Identify Desired Learning Outcomes".

Specify and Monitor the Teaching Process

Specification of the teaching process and ensuring it is carried out at a level of detail adequate to ensure its effectiveness is also implied by ISO 9000, as interpreted by Shutler and Crawford (1998). The requirement is that, at suitable points in the teaching process, monitoring of teaching and learning outcomes must take place. This is to ensure individual lecturers are not allowed to decide their own teaching methods, particularly if they are not effective. At UWI, the Centre for Excellence in Teaching and Learning offers the postgraduate Certificate in University Teaching and Learning and also runs training workshops for academic staff to enhance their pedagogical and andragogical skills. New members of academic staff, "including tutors, without teaching qualifications, are . . . now mandated within their hiring contracts" to enrol in Certificate in University Teaching and Learning (UWI 2010). New academic staff must do so within the first three years of employment (UWI 2010). What is required to move quality at UWI to the next level is for larger numbers of academic staff to more frequently avail themselves of the training provided by the Centre for Excellence in Teaching and Learning.

Assign Teaching Duties Based on Training or Experience and Identify Lecturers' Training Needs

Even when effective teaching methods are indicated, students may fail to learn if lecturers lack the necessary skill or training to use the methods specified. Consequently, ISO 9000 necessitates assignment of teaching duties based on lecturers' training or experience. The standard also requires identifying and providing for lecturers' training needs. Here again, the Centre for Excellence in Teaching and Learning plays an important role designing and offering training workshops to ensure that lecturers possess the required training and skill to employ particular teaching methods. However, it is reported that often it is the same members of academic staff who avail themselves of these opportunities, with the implication being that academic staff who do need to attend Centre for Excellence in Teaching and Learning training do not. Instituting a requirement that makes it mandatory for all members of academic staff, including senior members of academic staff, to expose themselves to ongoing training in pedagogical methods would be moving quality at UWI to the next level. This could very well have the effect of a reduction of student complaints about ineffective teaching methods as well as the organization and delivery of teaching generally.

Ensure Desired Learning Outcomes Are Achieved

Graduating students may not have learned what they expected to learn. Reasons for this could be either failure to assess them in some areas or methods of assessment being "incapable of distinguishing between satisfactory and unsatisfactory learning outcomes" (Shutler and Crawford 1998, 157). The standard, therefore, requires examination of students in ways that ensure achievement of the desired learning outcomes before the award of a degree certificate. The examination "should be calibrated against nationally recognised standards of attainment or, where no such standards exist, against standards specified by the university" (Shutler and Crawford 1998, 157). At UWI, the Centre works with teaching departments in developing learning outcomes as part of programme development. There are, however, departments with courses designed before this thrust. Ensuring that all courses offered have well-defined learning outcomes would be moving quality at the UWI to the next level. The university's introduction from the 2014–15 academic year of a revised grade point average scheme is reflective of the requirements of this standard. With the revised grade point average scheme, "grade descriptors" will be introduced to explain the level of skills and knowledge represented by each letter grade. It is expected that this will lead to "an approach that is more transparent and objective, and designed to give a clearer picture of the level of competencies achieved by each student" (UWI 2012, 1).

Give Clearly Defined Responsibility and Authority to Administrators of the Teaching System

If administrators who know what is going wrong are not given the authority to correct what is not working, students who have not learned what they have been taught may nonetheless graduate. Administrators of the teaching system must, therefore, be given "clearly defined responsibility and authority to make the necessary changes" (Shutler and Crawford 1998, 157). At the university, deans and heads of departments are the chief administrators of the teaching system and, together with lecturers, are well placed to know what is or is not working in the classroom and to take corrective action. Where broad policy issues that impact on teaching and learning are involved, however, decision making occurs at higher organizational levels such as Faculty Board, Academic Board and regional boards of governance such as Board for Undergraduate Studies and Board for Graduate Studies and Research. These institutional processes may not always facilitate the exercise of responsibility and authority by the immediate administrators of the teaching system. Some review of the limitations of the functioning of these administrative mechanisms would indicate whether any change is needed in keeping with this requirement of ISO 9000 and, if so, the nature of the changes warranted. Such deliberations must take into account the requirement of the standard that "educational management tasks should be assigned on the basis of appropriate training or experience, and the training needs of management should be identified and provided for" (157).

The provisions of ISO 9000 as applied to higher education require universities to understand what students are interested in learning and to adapt syllabi accordingly; employ effective teaching methods that are delivered by lecturers trained to use them; use assessment strategies to guide selection of students; and ensure graduates have achieved desired learning outcomes. Further, the standard requires commitment of institutional authority to implement these provisions.

In respect of this latter requirement specifying the need for commitment of institutional authority, specifically the features of ISO 9001:2000, QMS process approach seems useful in the case of UWI. This standard provides for the clear identification of management responsibility for the QMS; using the quality policy to guide decision making; ensuring a customer focus in the educational organization; and the provision for communication processes vertically and horizontally to share information about the QMS.

Selected Quality Models

The intention of this chapter is not to suggest that UWI should seek ISO certification but rather that the guidelines that constitute ISO 9000 and ISO 9001:2000 are likely to be useful components for an integrated and enhanced UWI QMS that would, at the same time, benefit from components of other approaches to quality, particularly suited to meeting specific needs of an educational institution. As indicated earlier, this flexibility is in keeping with the spirit of ISO 9001:2000, in particular. Accordingly, the chapter goes on to consider aspects of selected quality models that can help to move quality at UWI to the next level.

Selected quality models that appear useful for the improvement of academic and administrative quality, from the perspective of addressing organizational and programme weaknesses indicated in the student feedback, and moving quality at UWI up to the next level include: the "Plan–Do–Check–Act" model; Root Cause Analysis and the Excellence in Higher Education model. In the feedback they provided, students called for "proper organisation"; "more efficiency"; "more organization information centrally located"; "better organization of classes"; "more effort in organization" (patients, labs, classes, clinic); and "better communication among lectures and with students".

At the university, the PDCA model could be used to improve organization and delivery of teaching to ensure effective learning. Useful features of the PDCA model include a continuous improvement cycle, with the identification of areas for improvement using a strategic planning process, informed by the application of best practice. This latter component of PDCA is particularly important, lending to the PDCA model a transformative potential. It is not that UWI does not engage in strategic planning, for it does. Rather it is not at all evident that operational plans developed and implemented according to the strategic plan are sufficiently informed by best practice. The model also incorporates anticipated outcomes and criteria for identifying success. The PDCA cycle can be applied to the process approach embodied in the ISO 9001 standard and this adds to its utility.

Root Cause Analysis is another quality model to be considered to move quality at the UWI to the next level. A root cause is the underlying breakdown or failure of a process which, when resolved, prevents the problem from recurring (Sherwin 2011). Student feedback discussed above suggests that there exists an underlying failure of processes related to the organization and delivery of some academic programmes at UWI. The adoption of Root Cause Analysis as an important component of quality improvement can help to move quality at the UWI up to the next level where such programmes are concerned. Objectivity and a thorough and disciplined approach are key characteristics of this quality model employed to determine the most probable underlying causes of problems and undesired events within an organization. The aim of using the model is to formulate and arrive at consensus regarding corrective actions that can at least mitigate, if not eliminate, causes of problems thus producing significant long-term performance improvement. This quality model makes available to the institution a systematic approach to get to the root of a problem to prevent it from recurring. The approach involves identifying the problem, defining the problem, understanding the problem, identifying the root cause, taking corrective action and monitoring the system. A questioning technique involving asking "Why?" at least five times is employed to get to the root of the problem.

The Excellence in Higher Education model, an adaptation of the Baldrige framework, is particularly useful as it employs the language of higher education, is adaptable to an institution's mission, and is applicable to a higher education institution's administrative service areas and traditional academic areas. Its seven categories of assessment are leadership, planning, external focus, measurement and knowledge utilization, faculty/staff and workplace focus, process effectiveness and outcomes, and achievements. These categories correspond to what are viewed as critical components of, as well as contributors to, excellence of an educational institution. The seven categories and their interactions define a system framework that can be used to conceptualize and analyse the workings, effectiveness, strengths and improvement needs of a higher education department, programme or institution. The Excellence in Higher Education model is useful in a workshop or retreat setting as an integrated self-assessment, priority-setting and action-planning tool. It can therefore perform an evaluative function in support of improving the quality of planning, organization and delivery of academic programmes.

Conclusion

Using components of the selected ISO standards and quality models discussed, with elements of the ISO standard as the mainstay, an eclectic quality model can be fashioned for an integrated approach to academic and administrative quality at UWI. This does not negate the continuing utility of the "fitness-for-purpose" approach that is applied to quality assurance reviews of academic programmes. Such an eclectic model can be used to guide planning and implementation of academic programmes and other activities and, along with the "fitness-for-purpose" approach, therefore, can ensure continuous improvement not only of teaching and learning but also of governance and management of quality and implementation of quality policy at the UWI.

Acknowledgements

The author acknowledges the kind assistance of Candice Hickson, research technician, Quality Assurance Unit, University of the West Indies, in the preparation of this chapter.

References

BSI Quality Assurance. 1995. *Guidance Notes on the Application of BS EN ISO 9000 for Quality Management Systems in Education and Training*. London: BSI Quality Assurance.
ISO (International Organization for Standardization). 2003. *Quality Management Systems: Guidelines for the Application of ISO 9001:2000 in Education*. (International Workshop Agreement.) Geneva: ISO.
Sherwin, J. 2011. "Contemporary Topics in Health Care: Root Cause Analysis". *PT in Motion* 3 (4): 26–31. http://web.ebscohost.com/ehost/pdfviewer/pdfviewer?sid=de90ef45-396f-45f5-aea2-b8245cec34c0%40sessionmgr13&vid=2&hid=11.
Shutler, M.E., and L.E.D. Crawford. 1998. "The Challenge of ISO 9000 Certification in Higher Education". *Quality Assurance in Education* 6 (3): 152–61. http://www.emeraldinsight.com/journals.htm?articleid=839609.

UWI. 2010. "Institutional Accreditation Self-Study Report". Office of the Principal, St Augustine, Trinidad.

———. 2012. "Introduction of a Revised GPA Scheme 2014/2015 Academic Year". Office of the Board for Undergraduate Studies, UWI Regional Headquarters, Mona, Jamaica.

15

Details, Details, Details
Administrative Personnel and Quality Assurance

JUNE WHEATLEY

SCENARIO

A prominent tertiary education institution in the Caribbean recently promoted a member of its staff, who has been in its employ for over twenty years, to the position of quality assurance officer. This was a new position created to enhance the quality and standards of the institution's programmes, as well as to meet the standards of the national accrediting body.

The officer's main responsibility was to set up a quality assurance system within the institution. This entailed, among other things, writing policies and procedures to ensure that the programmes offered by the institution were aligned to the institution's vision and mission, as well as meet national and international standards.

The officer was expected to undertake this new assignment without the necessary physical and human resources. Firstly, there was no provision of office space for the officer; secondly and equally significant, no administrative support was provided with this important post. The officer explained that it is with much difficulty that she functions without administrative support. She spent a lot of her time researching, arranging and rearranging meetings, requesting data – all the functions administrative support would have carried out; this would allow the officer to concentrate on her core role.

The importance of efficient and effective administrative professionals in an organization, though obvious, is oftentimes undervalued and overlooked, as the scenario above demonstrates. Indeed, the work of a quality assurance office depends significantly on a cadre of trained administrators, who take the lead on the detailed processes in a quality assurance review or evaluation. In the big picture approach to visioning quality assurance, the administrative transactions are taken for granted and often made invisible. Yet, they are an essential component, as the quality assurance experience at the University of the West Indies (UWI) has shown. At the same time, very little has been written on the role of administrative support personnel beyond what appears in the usual office practice or office procedures texts (for example, Fulton-Calkins et al. 2011; Stroman et al. 2012). Perhaps the administrative support function has been too often collapsed into the swiftly disappearing secretarial role. (The 2011 findings of the IAAP survey suggest that this is not the case, however.) The International Association of Administrative Professionals (IAAP), in its 2013 benchmarking survey, highlights administrative professionals as having "a vital role in the success of their employers, yet their value is little understood" (1). The IAAP calls for quantifiable data not only to help increase the effectiveness and efficiency of the work of administrative professionals but how they are trained and managed.

This chapter takes seriously the IAAP charge to provide quantifiable data to help increase the effectiveness and efficiency of the work of administrative professionals. It does so specifically by looking at the work of administrative professionals in the realm of quality assurance in higher education in the Caribbean. It explores the quality assurance process from the perspective of the administrative support staff drawing on experiences of such staff from the regional University of the West Indies; the University of Technology, Jamaica; Northern Caribbean University (Jamaica); Excelsior Community College (Jamaica); and The Mico University College (Jamaica). In so doing, it calls for a new appreciation of the value of administrative support as significant partners in the quality assurance endeavour (indeed, in higher education generally) and outline ways of deepening the professional engagement and development of such staff outlines. (In this study, the terms administrator, administrative personnel, administrative assistant and administrative support staff will be used interchangeably.) The major outcome of this chapter is to prepare a checklist for administrative professionals involved in the quality assurance process, based on the UWI, Mona, experience. This generic checklist can be adapted to suit the needs of individual quality assurance offices.

Quality Assurance at the UWI

The UWI experience is the model against which the research in this chapter is measured. UWI is the premier tertiary education institution in the Commonwealth Caribbean and has had the longest tradition of a formal internal quality assurance system involving review and evaluation (audits) of all its academic programmes and several aspects of the learning environment, such as student services. The robust internal process was initiated in response to the 1994 Chancellor's Commission on Governance, which called for a fully integrated system of quality assurance reviews and audits (Whiteley 2002, 262). The Quality Assurance Unit (QAU) in the Office of the Board for Undergraduate Studies has the responsibility for the academic quality assurance system. The university through the Office of the Board for Undergraduate Studies, prior to the formation of the QAU, and later through the QAU which was formed in 2001, has a carefully worked-out process for conducting reviews and evaluations, which has been documented and refined through the years based on international best practice, experiences gained as well as recommendations from the various external review teams.

The administrative professionals in the QAU on the different campuses of the UWI have received commendations from various review teams over the years. For example, the 2005 review team for the Department of Literatures in English, Mona stated: "Most of all, we want to convey our warmest thanks to . . . Administrative Secretary in the Quality Assurance Unit [Mona], for her matchless efficiency, and organization skills, and cheerfulness" (Quality Assurance Unit 2005, 5). Similarly, the 2011 review team for the Social Work Unit at the St Augustine campus expressed their indebtedness to the senior programme officer and the administrative assistant "for their highly professional and competent organization of the Review, both during the preparation stages and the actual Review" (Quality Assurance Unit 2011b, 5).

The QAU itself underwent an external review in 2010 and is currently implementing improvements in tandem with the strategic objectives of the University. The review team in their feedback to the Mona QAU stated that "your organization . . . of the lead up to the event [review team visit] is excellent" (Quality Assurance Unit 2011a, 1). This statement was made in reference to the administrative arrangements made for the team prior to the site visit. The team went on to commend the involvement of the administrative assistant at Mona in the quality assurance process and recommended an expansion of the job functions of other administrative assistants in a similar way. The team also recommended that the administrative staff meet regularly in the same way that quality assurance officers currently do.

In their overall report, the team noted that

> QAU officers work effectively as a team across campuses and this has helped to make the most of the limited human resources in these Units. They are also well supported by clerical staff, and the officers of the central OBUS unit at Mona. The job functions of the Administrative Assistants (AA) in the delivery and management of programme activities of the QAU include a wide variety of activities which are summarized in the SAR (SAR, 2011:52). (Quality Assurance Unit 2011a, 21)

Again, the team highlighted the centrality of administrative ("clerical") support to the work of the QAU. They also noted the variety of tasks involved in the administrative role.

In addition to the internal quality assurance mechanisms, all the campuses of the university are accredited by the relevant national accrediting bodies, University Council of Jamaica (Mona), Accreditation Council of Trinidad and Tobago (St Augustine) and Barbados Accreditation Council (Open and Cave Hill). Professional programmes such as medicine and engineering are accredited by international and regional professional agencies. The accreditation teams that visited the four campuses of UWI over the past two to three years have expressed high commendations of the system. As a matter of fact, even before that UWI has been commended in different international fora, such as conferences, by peers who acknowledged that UWI was way ahead of other international institutions in terms of its system of quality assurance. For example, at the 2002 qualitative assurance conference in Dublin, participants lauded the effort of the UWI in this regard (Anthony Perry, interview by author, September 12, 2013). The university quality assurance system has also been regarded as a benchmark by other institutions in the region. The unit has undertaken successful consultancies and workshops on quality assurance matters for diverse institutions such as the University of Guyana (2011), the Ministry of Education in Suriname (2006) and the Northern Caribbean University (2012), among others.

The Importance of Administrative Support

Sandra Richards, senior programme officer in the UWI Quality Assurance Unit with responsibility for graduate studies and research, sums up the work of the quality professional as follows:

> The role of the quality assurance professional is, in short, to ensure that appropriate and effective measures for establishing, monitoring and maintaining quality are in place and serving the intended purpose. However, it is no secret that adhering to the measures that ensure quality can be viewed as intrusive and overwhelming. Quality Assurance Officers bring a myriad of skills and knowledge that make quality requirements more manageable for stakeholders. Nonetheless, it is important to note that this function is significantly strengthened with the *right administrative support*. (Richards, email to author, September 25, 2013; emphasis added)

Richards is able to draw on the experience of having to initially operate without dedicated administrative support. With this unique insight into how these two circumstances impacted on her work, she argues that the function of administrative assistant extends way beyond being able to generate, store and retrieve documents and data. The administrator is completely aware of the institutional infrastructure and is often the conduit between the realities of key players, systems and intended objectives. The administrator will hold key information as well as access. S/he will be aware of the protocols and gatekeepers that can progress or prevent necessary processes. In my experience, the administrator is an essential colleague who works closely with the Officer as the somewhat invisible, yet critically astute, hinge-pin frequently employing tact and diplomacy to ensure that objectives are reached efficiently and effectively. The administrative assistant brings not only expertise but insight as s/he performs the vital role that underpins the work of quality assurance Officers. (Richards, email to author, September 25, 2013)

Richards identifies several key roles of the administrative support, included among these are acting as hinge-pin (perhaps linchpin) to ensure objectives are reached, holding key information and providing access to that information, and knowledge of key systems, personnel, and infrastructure. Marketing and advertising consultant and blogger at recruitingblogs.com, Theresa Boruta (2011) concurs with Richards's position when she argues that administrative assistants "act as the lifeblood for most businesses". Boruta rejects "[the] common misconception that it's an *easy job* that anyone can do" (2011; emphasis added). Explaining how seminal the administrative assistant is, she notes that "there are many very specific qualities that differentiate a good administrative assistant from a great one". She lists six qualities that great administrators should have. These include communication skills, organizational skills, time management skills, dependability and reliability, confidentiality, and customer or client service orientation. These qualities will be discussed later on in the chapter.

Administrative Transactions in the Quality Assurance Process

At UWI, the administrative transactions in the quality assurance process are many and varied, as noted by the review team of the Quality Assurance Unit (2011a). Many of these have to be carried out simultaneously in order to achieve high-quality outcomes. Two major quality processes in higher education institutions are accreditation and internal quality assurance reviews. Both the accreditation and review process require detailed administrative activities ranging from researching information on the recommended reviewers to ensure that they are suitable for the tasks, to the purchasing of airline tickets, booking accommodations and preparing subsistence for review team members. Other activities include booking meeting rooms, preparing schedules, organizing meetings, organizing catering, preparing folders for team members, collecting and collating documentation, and liaising with administrative personnel in the department being reviewed to ensure documentation, including the self-assessment report, is prepared on time. While the team is on the ground, the administrative support has to keep abreast of the day-to-day and hour-to-hour activities. Administrative personnel also proofread and edit reports produced by the team. If any of these activities, for example, the purchasing of airline tickets and booking accommodations, are not done in a timely manner, the institution may be burdened with additional cost. As well, the quality assurance review process will be less effective. Failure to monitor these processes closely to ensure that things happen how and when they ought to, will result in the quality assurance process being compromised and the Quality Assurance Unit and institution seeming inefficient and incompetent.

Method

The case study method was chosen to do this research. Robert K. Yin, one of the leading proponents of the case study strategy, defines a case study as an empirical inquiry that investigates a contemporary phenomenon within its real-life context (1994, 13). Also worthy of note is Gerring's definition, which states that a case study is "an intensive study of a single unit with an aim to generalize across a larger set of units". He argues that case studies rely on the same sort of covariational evidence utilized in non-case study research and thus is correctly understood as a particular way of defining cases and not a way of analysing cases or a way of modelling causal relations (Gerring 2004, 341). Administrative professionals who participate in quality assurance processes in their institutions were chosen for this empirical inquiry. Within this case study, the focus group discussion and survey methods were used in order to ascertain the various activities undertaken by the administrative support function in quality assurance, as well as the skill set needed for effective functioning. The survey instrument was followed up with telephone interviews. Silverman (2006) defines focus groups as "group discussions usually based on visual or verbal stimuli provided by a researcher" (401). Leedy and Ormrod (2005) note that focus groups are especially useful when "time is limited, people feel more comfortable talking in a group than alone" and "interaction among participants may be more informative than individually conducted interviews" (146). The focus group was used as it is considered the most suitable method of collecting qualitative data which would be difficult to acquire using other methods. Participants in focus group discussions, accommodated in a comfortable setting, usually feel safe and secure to speak honestly and frankly. The researcher followed Leedy and Ormrod's guidelines for focus group discussions by ensuring that "everyone in the group [had] a chance to answer each question" (149). The survey was also administered in order to collect data from administrative personnel in Jamaica but outside of Kingston, as well as in the Caribbean.

The study suffered from several limitations. The study was confined to administrative personnel, who participate in quality assurance processes in their institutions; this was a small group as not many higher education institutions in the region have dedicated quality assurance offices. However, personnel with identifiable academic quality-related roles were targeted. One person with the title quality assurance officer was included as, being without dedicated administrative support, she functioned as such.

Three of the seven confirmed administrative personnel attended the focus group discussion, which was held May 15, 2013. In addition, questionnaires were emailed to twelve administrative personnel in Jamaica, Trinidad and Tobago, and Barbados. Four persons returned completed questionnaires after several email and telephone follow-ups. These include two from Jamaica and two from Barbados. The total number of participants was seven – one male and six females. Five Jamaicans and two Barbadians participated in the research from institutions, including the Council of Community Colleges of Jamaica; Excelsior Community College; Northern Caribbean University; the University of Technology, Jamaica; the University of the West Indies (Cave Hill and Open Campuses); and the Mico University College. In order to protect the identity of participants and the confidentiality of their responses, they were labelled P1 to P7. Given the relatively small size of the sample, the results are considered to be indicative rather than representational and therefore generalizations cannot be drawn from this study; nonetheless, but it can serve as a launch pad for further research. While the researcher used the IAAP 2013 Benchmarking Survey to illuminate her findings, a true comparison of both studies could not be made as the populations studied are different. The principal researcher is herself a senior administrative assistant in the QAU at the UWI, serving the Mona campus, and so the research undertaken is directly related to her own function. The researcher's own experiences will be taken into account in the process and so the study is by no means a disinterested one.

Demographics

Most of the respondents were within the 26–35 age range. Of the seven participants, one was in the 18–25 age range; five were within the 26–35 age range, while one participant was aged over 55. This is in contrast to the IAAP survey, where 50 per cent of the respondents were in the 41–55 age range. The disparity in age range between both studies may be due to the fact that the Jamaica has a younger population, while the US population "has a substantially faster growth rates at older ages" (Howden and Meyer 2011, 2). Job titles ranged from stenographer clerk to administrative assistant, quality assurance officer and assistant examinations officer (who was an administrative assistant before she was promoted two weeks before the focus group discussion).

Respondents have been in their post between two weeks and three years. However, at least three persons were working in their organization for seven years and more before they were promoted to their present position.

Academic Qualifications

The 2013 IAAP survey notes that "one of the qualities that defines a career-minded administrative professional is the active pursuit of their own career education and professional development" (4). All the administrators (100%) had at least an undergraduate degree. This is in contrast to the qualifications of respondents in the IAAP study which reports that while 90 per cent of respondents had some sort of postsecondary education, only 45 per cent had at least an associate degree. Respondents' areas of specialization in the study include public sector management, psychology, management studies, education and social work. These were highly qualified and highly motivated individuals. All respondents reported that their studies helped them to perform efficiently in their jobs. Clearly, administrative professionals do not necessarily require formal training in quality assurance in order to perform satisfactorily. None of the administrative support had formal training in quality assurance, quality enhancement or accreditation. As the study found, their broad educational background and experience on the job provide the platform for them to perform.

Findings

As expected, the administrative work in quality assurance in higher education is highly transactional. All administrators were integrally involved in the quality assurance processes at their institution. These include requesting documentation from departments and ensuring that the departments document procedures. Data production, documentation and analysis are recognized as key elements in the quality assurance process. P6, who has been in her present position for three years, but has been with her institution for seven years, noted that she had to ensure that all lecturers complete a faculty module assessment report. This report form is designed to assist in examining the module (assessment, delivery, content, and so on) with the aim of ensuring improvement to the module at the next offering. Information gleaned from the faculty module assessment report, for example, may help to determine whether the curriculum should be modified. Administrators reported that they had to exercise their people and communication skills significantly in order to accomplish these tasks. For example, P4, who has been in her job for two years, asserts that "my degree in psychology helps me to negotiate with departments that are reluctant to document their procedures, and convince them to prepare procedures manuals in furtherance of the quality mandate". She was able to accomplish this as she had learned about the different personality types and reactions in different situations. Another (P3) cited an example of her institution where the quality assurance officer was perceived as arrogant and "high-handed". The academic staff of the institution would often not cooperate in submitting the relevant documents requested in light of their perception of their colleague. It was the administrative support, who in employing her people skills, ensured that the documentation was eventually submitted.

Another aspect of the process that some administrators are engaged in was the organization of internal quality assurance reviews and evaluations. These administrators reported having to pay keen attention to details in the planning, organization and execution of these events. As detailed in the UWI experience above, organizing an internal review requires booking and purchasing airline tickets for review team members, booking accommodation, preparing schedules, arranging meetings, organizing meals, arranging ground transportation and a host of other activities. At the same time, it requires working with colleagues internally, specifically the department being reviewed, to get documentation needed for the team, request names and contact information for the different stakeholders in a timely manner. This requires being extremely meticulous and paying keen attention to details. Time management and multitasking are also necessary skills in this process. For instance, P5, a male who was the newest member to his institution's quality assurance office, having joined only eleven months earlier, posits that "the ability to multi-task, plan and coordinate are essential to preparing for reviews and evaluations". He notes that he "coordinates QA activities for different sites across the Caribbean", which includes "coordinating travel arrangements for review team members and create travel and review schedules for distribution to the site which is hosting the review". He also states that "I have to ensure that the completed self-study is forwarded to the different stakeholders. As the review approaches, I would prepare packages for the review team which include travel and review schedules and the Self-Assessment Report."

Understandings of Quality Assurance

Most of the persons interviewed had a similar understanding of what is quality assurance. The majority see quality assurance as the implementation of systems to improve and enhance established standards of products and services. Others view quality assurance as the policies and procedures in place to safeguard academic standards and promote learning opportunities of acceptable quality for students. P4, for example, noted that at her institution the emphasis is on quality management and, therefore, quality assurance for her is the "improvement of processes so that the end product has minimal or no defects". P7, who has been with her institution for eleven years, noted that quality assurance is "fitness-for-purpose" as "we have a responsibility to stakeholders to ensure that we are actually doing what we say we do". P7's understanding of quality assurance is similar to the model which guides quality assurance in higher education generally. The fitness-for-purpose model presumes that "quality is judged in terms of the extent to which service meets its stated purpose" (OBUS 2000/2001, 167). It is clear from the responses that administrative personnel have a good understanding of how quality assurance is generally and specifically understood in their institutions.

How Administrative Support See Their Role in the Quality Assurance Process

Most administrators see themselves as being part of a team, and consider themselves partners in the quality assurance process. They consider their role as vital in the scheme of things. P5 explained his role as "providing logistical support to the QA officer, which entails assisting with the coordination and planning of the various quality assurance processes". While P6 stated, "I see my role as an integral part of the team, ensuring that all self-studies for the Faculty are complete and that all the courses of study are prepared for both local and international accreditation."

Attributes of the Effective Administrator

"One must have great organizational skills, be able to multi-task, have an eye for detail, be a people person as well as have great communication skills" (P7). P7's statement sums up the key requirements most administrative support believe they need to have to be functioning optimally in their jobs. While it was agreed that academic qualifications are important in the quality assurance process, respondents felt that people skills are equally important.

Boruta (2011) notes that communication (interpersonal relationship and listening) skills are possibly the top desired qualities every employer looks for in an administrative assistant. Effective administrative personnel have to know how to prioritize and so Boruta believes organizational and time management skills are two very important attributes that one should possess. Much responsibility is placed on administrative personnel and therefore they are expected to be dependable and reliable. They are expected to go above and beyond the call of duty in urgent as well as daily situations. Dependability and reliability are desirable and important attributes. For example, during the week of a quality review, the administrative support is expected to be not just on time for work, but to be very early, arriving before the team members to ensure that things are in place and to smooth out any problems that may arise.

People skills or, as Boruta calls it, customer service orientation (2011) is an important attribute for administrative support. Boruta notes that while customer service orientation is desirable in all professions, it is considered to be an especially important attribute for administrative professionals. She argues that "no matter how skilled an assistant is in every other area, if he/she is not personally invested in helping others and guaranteeing satisfaction, the company will not benefit".

Stulz, Shumack and Fulton-Calkins (2013) offer a more comprehensive list than Boruta (2011) of attributes that administrators should possess. The experiences of the participants in this study correlate with both Stulz, Shumack and Fulton-Calkins (2013) and Boruta. Stulz, Shumack and Fulton-Calkins listed twelve skills needed in all administrative professional positions. These include communication (listening, reading, verbal presentation and writing), interpersonal relations, time management, critical thinking, decision making, creative thinking, teamwork, technology, leadership, stress management, problem solving and customer focus. Table 15.1 compares the attributes identified by Stulz et al., Boruta and the participants in the survey and focus group. Of course, there is significant overlap among the three sources. The focus group and those surveyed also identified most of the same skills listed by Stulz et al. and Boruta as necessary in their job functions. While the participants identified many of the attributes required, teamwork, creative thinking and decision making were not among attributes identified, however. Participants spoke of the need to be "analytical" and have "people skills". The participants' attribute "analytical skills" correlates with Stulz's "critical thinking skills". Similarly, the participants "people skills" is akin to Boruta's "customer or client service orientation".

Ironically, none of the participants listed dependability and reliability, stress management and confidentiality as attributes required. It is possible that these attributes are taken for granted by the participants, however. While it can be concluded that the administrative professionals who participated in this study are aware of and in tune with the requirements of their job, for the most part, there are identifiable gaps. Attributes such as confidentiality and creative thinking are intrinsic qualities for which administrators have to take responsibility and work on these qualities themselves.

Resources for Administrative Professionals

There is no doubt that the role of the administrative professional is crucial to the efficient operations of the QAU. However, from the discussion held, it is clear that

Table 15.1. Skills needed by effective administrators

Attributes identified by administrators	Attributes from Stulz et al.	Attributes from Boruta
Communication skills (P3; P6; P7)	Communication skills	Communication skills
Office management skills (P4)	Confidentiality	Organizational skills
People skills (P3; P7)	Interpersonal relations	Customer or client service orientation
Time management (P6)	Time management	Time management
Analytical skills (P4; P6)	Critical thinking	Dependability and reliability
Eye for details (P7)	Decision making	Confidentiality
Project management	Creative thinking	
	Teamwork	
Computer skills (P6) and word processing (P4)	Technology	
Taking own initiative	Leadership skills	
Critical thinking (P3)	Stress management	
Organizational skills (P3; P7)	Problem solving	
	Customer focus	
Interview skills (P4)		
Multi-tasking (P5; P7)		
Planning (P5)		
Coordination (P5)		

Sources: Boruta (2011) and Stulz, Shumack, and Fulton-Calkins (2013).

many institutions have neglected to see this as a priority area thereby provide adequate resources. Fifty-seven per cent of the respondents reported extreme levels of frustration with the lack of basic equipment such as scanners, printers and adequate computers to carry out their job functions. P3, for example, reported that her computer takes over ten minutes to boot up each morning; she does not have access to a printer and could do with better technical support. P2, who has been an administrative assistant two and half years prior to being promoted two weeks before the focus group discussion, noted that "resources are also very important; unavailability of resources hampers your ability to perform". Not surprising, the IAAP reports that "a faster, more powerful computer is the *number one* item on the technology wish list for administrative professionals. That's followed by a tablet, additional or upgraded software, smartphone and a larger computer monitor." The IAAP reports that "only one-third of the respondents manage office calendars on external devices such as smartphones or tablets" and that "this is low given the proliferation of mobile tools among other groups of office professionals". The IAAP noted its concern that "employers may be creating a kind of digital divide within their office teams by not providing mobile tools". The lack of resources seems a major challenge for most administrative personnel in their institutions. This, they report, is a significant hindrance to the efficient performance of their job. Employers should therefore endeavour to provide adequate resources for administrative professionals to carry out their tasks.

In the focus group discussion, another area of concern for administrative personnel in the quality assurance process is professional development. While administrators agree that they do not need to know all the areas in quality assurance to carry out their task, at least one person felt that she would like to receive some training in the area. P2 noted that "it is important to keep up to date with best practices in quality assurance. Administrative personnel should therefore be able to attend conferences, workshops and seminars in order to stay current in their area."

Administrators raised concern that, while their academic counterparts are given support for professional development, administrative personnel are often neglected. For example, P3 asserts that at her institution staff development "is mostly focussed on academic and ancillary staff, while administrative staff is left in the middle". They usually have to seek professional development on their own initiative and with their own resources. Contrastingly, the IAAP study shows that "approximately two-thirds of respondents say their employers provide in-house training" and "three-fourths [say] employers will pay for administrative professionals training provided by outside vendors". The IAAP was concerned, however, that the "data was less encouraging when it comes to the amount of professional development provided to the survey respondent". The IAAP argues that "10 hours of training is simply too little to be adequate for an administrative professional in the contemporary office" and stated that "this is an area where improvements can and should be made" (IAAP 2013, 4).

P2 argues that "it would be a motivating factor if my institutions would invest in my professional development by sending me to seminars, workshops and conferences". This would benefit not only the individuals themselves, but also the institution as well – a motivated and satisfied individual is a more productive individual. In fact, one of the classical management science theorists, Frederick Taylor, recognizing the importance of the need for organizations to facilitate employees training, emphasized that "Once the right man [person] had been employed for the right job, management then had the responsibility of properly training that person. The process of training and development ought (in differing degrees as called for by the particular situation) to apply to *all* employees, whether in the 'science of shovelling' or in the development of the most senior executive" (Taylor 1911, quoted in Morden 2004, 13).

Respondents were asked whether there was any documentation that outlines their job function. Two respondents (P3 and P4) noted that the only document available was their job descriptions.

P5 and P7 cited quality assurance modules in their institutions, which help to some extent to describe the work of quality assurance at their institutions. However, the modules do not outline their job functions. P4, who has been an administrator in the Quality Management and Assessment Department at her institution for two years, noted that while her job description is currently the only documentation that exists that details her work in quality assurance, she plans to change that situation soon, as she would be writing a procedures manual for her job function. This is a very commendable initiative in which administrative professionals involved in the quality assurance process could be engaged.

Deepening the Professional Engagement of Administrative Support in the Quality Assurance Process

"Managers should consult their administrative support for solutions to problems and seek our advice before revising system" (P3).

Administrators ensure that all the minute details in the quality assurance processes are attended to. They recognize that failure to address these details can cause their institutions major embarrassment as well as additional expenses. More importantly, this can compromise the process resulting in a poor or inaccurate report (in the case of a review) or an incomplete process.

In preparing for a quality assurance review, there are certain details that need to be taken into consideration. One of the tasks that the administrator does is the organization of travel and accommodation for the review team. Although this seems a fairly straightforward activity, it can be complex. One has to be analytical and must exercise patience in dealing with individuals. Sometimes, requests are made that are not in line with the institution's policy. The administrative professional has to be firm but pleasant in his or her response. For example, a team member may request that you purchase a first-class ticket for his travel or ask that your institution cover the cost of the ticket for a family member out of his or her honorarium. The administrative support has to be able to say "no" to these requests in a professional and courteous manner, while explaining the institution's policy.

Time management and organizational know-how are important skills administrative professionals need to possess. One has to be able to multitask to meet different deadlines and to be on top of the different activities happening at once. It is always best to work with a checklist, with timelines (see appendix 15.1).

Administrative professionals are highly competent individuals. However, it appears that only few managers seek to deepen the professional engagement of these individuals. While statistics for the Caribbean are not readily available, in an American Management Association (AMA) survey (2002), 76 per cent of administrative professionals reported that they had more responsibilities over the prior year. However, only 2 per cent felt that they would assume their manager's role should he or she vacate the job. The majority of the respondents in the American Management Association survey (66 per cent) felt they would remain in their current position. It would appear therefore that while the administrative professional is experiencing job expansion, they are not really experiencing job enrichment. So, for example, an administrative professional might have started out supporting one executive, and later one or two other executives are added to his or her portfolio. She or he may have more responsibilities (job expansion) but not necessarily job enrichment as the person may be doing the same thing (two/three times). Similarly, the IAAP noted that in 2009 "administrative professionals experienced a marked increase in the number of managers and/or executives they were expected to support" (IAAP 2013, 2). The American Management Association survey (2002) noted that in spite of important gains in responsibilities, a significant proportion of administrative professionals still do not believe they receive the respect they deserve from their co-workers. This seems to differ from the 2013 IAAP survey findings, however, where the results "suggest that the diversity and prestige of positions held by administrative professionals continue to increase".

Nonetheless, in an effort to enhance the quality assurance processes, managers should endeavour to deepen the professional engagement with administrative support. This can be done through job enrichment. For example, at UWI, Mona, the quality assurance officer includes her administrative support as part of the team in the quality evaluation of franchised programmes.

Conclusion

From the foregoing discussion, it is clear that the administrative professionals who participated in this study are highly motivated and competent individuals, who have taken on and are willing to take on more responsibilities in an ever-changing work environment. These administrators are aware of the skills and competencies they need to perform effectively in the work environment. They identified most of the attributes as determined by Stulz, Shumack and Fulton-Calkins (2013) and Boruta (2011), such as communication and organization skills. However, there were some attributes that these writers considered important, but were not readily identified by the administrators who participated in the study. These may be areas that administrative professionals could target for training.

Three major findings of this research, which may not be generalizable, but are worthy of consideration, are:

1. Administrators are not provided with adequate resources to carry out their work.
2. While administrative support update their skills and acquire training on their own initiative, institutions do not provide administrative professionals with adequate opportunities and funding to update their skills and competencies.
3. Administrative professionals were not provided with manuals specific to their work.

As such, appendix 15.1 provides a checklist that can be used as a guide for administrative professionals in the quality assurance process.

It is important for institutions to dialogue with administrative professionals to see where the gaps are and try to close them. After all, the administrative professional is an important part of the partnership in the quality assurance process.

Appendix 15.1. Checklist 1: Preparing for a QA activity

Department: _____

Head of Department: _____ Tel: _____

Name of Administrative Support to HOD: _____ Tel: _____

Item	Time frame/ Responsible party	Notes/Date completed
Draft list of QA activities for the year	Beginning of each academic year	
Advise relevant stakeholders		
Prepare schedule for each QA activity		
Arrange meeting with HOD to conduct orientation session		
Conduct orientation session with relevant stakeholders		
Research names of recommended team members (review team)		
Contact and confirm review team		
Arrange flights, hotel, ground transportation, subsistence and honorarium for review team		
Receive self-assessment report (SAR) from department		
Review SAR and give feedback to department		
Send SAR to review team		
Draft schedule of activities for review		
Contact and confirm meetings with stakeholders (list all the stakeholders)		
Supervise review		
Receive draft report		
Send report for factual verification		
Follow up on issues of concern		
Proofread/edit and finalize report		
Circulate report to relevant personnel (list names of people)		

Note: This checklist may be adapted to suit the particular QA activity.

References

AMA (American Management Association). 2002. *Fall Administrative Professionals Survey*. New York: AMA.

Boruta, Theresa. 2011. "Great Qualities Every Administrative Assistant Needs". http://www.recruitingblogs.com/profiles/blogs/great-qualities-every.

Fulton-Calkins, Patsy J., Dianne Rankin and Kellie Shumack. 2011. *The Administrative Professional: Technology and Procedures*. 14th ed. Independence, KY: South-Western Cengage Learning.

Gerring, John. 2004. "What Is a Case Study Good For?" *American Political Science Review* 8 (2): 341–54.

Howden, Lindsay M., and Julie A. Meyer. 2011. *Age and Sex Composition: 2010*. 2010 Census Briefs, United States Census Bureau. http://www.census.gov/prod/cen2010/briefs/c2010br-03.pdf.

IAAP (International Association of Administrative Professionals). 2013. *2013 Benchmarking Survey*. http://www.iaap-hq.org/press-release/2013-iaap-benchmarking-report.

Leedy, Paul D., and Jeanne Ellis Ormrod. 2005. *Practical Research: Planning and Design*. 8th ed. New York: Pearson Education.

Morden, Tony. 2004. *Principles of Management*. Burlington, VT: Ashgate.

OBUS (Office of the Board for Undergraduate Studies). 2000/2001. *Quality Assurance Strategy: The System in Action*. 1st ed. Kingston: OBUS, University of the West Indies.

Quality Assurance Unit. 2005. "Report of the Review of the Department of Literatures in English". Faculty of Humanities and Education, University of the West Indies, Mona, Jamaica.

———. 2011a. "Report of the Review of the Quality Assurance Unit". University of the West Indies, Mona, Jamaica.

———. 2011b. "Report of the Review of the Social Work Unit". Faculty of Social Sciences, University of the West Indies, St Augustine, Trinidad and Tobago.

Silverman, David. 2006. *Interpreting Qualitative Data*. 3rd ed. London: Sage.

Stroman, James, Kevin Wilson and Jennifer Wauson. 2012. *Administrative Assistant's and Secretary's Handbook*. 5th ed. New York: AMA.

Stulz, Karin M., K. Shumack and P. Fulton-Calkins. 2013. *Procedures and Theory for Administrative Professionals*. 7th ed. Mason, OH: South-Western, Cengage Learning.

Whiteley, Peter. 2002. "Quality Assurance in Selected Caribbean Universities". In *Adult Education in Caribbean Universities*, edited by Ian Austin and Christine Marrett. Kingston: UNESCO.

Yin, Robert K. 1994. *Case Study Research: Design and Methods*. 2nd ed. Thousand Oaks, CA: Sage.

16 | Transforming Higher Education in the Caribbean
The Total Quality Management/ Service Quality Model

EDUARDO R. ALI

Higher education institutions (HEIs) have been described as autonomous, differentiated professional bureaucracies that respond slowly to change and innovation (Jensen 2010; Mainardes, Alves and Raposo 2011). To bring about change, HEIs learn how to adapt their internal organizational systems when triggered by external factors through single- and double-loop learning approaches (Tossey 2003; Morgan 2006). Double-loop learning is usually encountered during transformational change. In double-loop learning, the organization questions the validity of operating norms and thinks intelligently about how and why it operates in specific ways thereby revisiting or introducing strategic changes (O'Donoghue and Clarke 2010). It is here where organizational learning for transformation exists.

Self-evaluation has been traditionally adopted as a double-loop learning method for guaranteeing stakeholder satisfaction and making recommendations for change (Pedder and MacBeath 2008). While quality assurance evaluations are usually performed in organizations to identify recommendations for change, they do not usually ensure the total management of these changes for organizational effectiveness as achieved during quality management. There are many models for quality management in higher education. Some examples include total quality management (TQM), European Foundation for Quality Management excellence model, balanced scorecard, Malcolm Baldridge performance excellence/quality award, ISO 9000 and business process re-engineering, and so on (Becket and

Brookes 2008). A good example of double-loop learning is found in TQM, which marries institutional research, leadership decision making, quality assurance and quality improvement (Neefe 2001; Agasisti and Bonomi 2013; Dobrzanski and Roszak 2007). TQM has been cited as a strategic approach to organizational transformation because it supports customer satisfaction while managing continuous improvement across the institution (Hogg and Hogg 1995). Interestingly though, the literature has shown cases where TQM approaches have not been very successful in the developed and some developing jurisdictions (Al Tasseh 2013; Pineda 2013). Morgan cautions that 70 per cent of organizations fail with TQM because they try to model the practices within old bureaucracies and not redesign the organizations to accommodate TQM (Morgan 2006). Zabadi (2013) argued that resistance by faculty members to change, inability by management to commit to TQM approaches, lack of continuous training, limited experience with TQM and inability to use TQM tools are some reasons that work alongside poor organizational designs to make TQM problematic. Dounos and Bohoris (2007) identify process improvement as key to TQM approaches that can be ignored by HEIs. The literature shows that other quality management models can be synchronous with TQM (Becket and Brookes 2008).

Notwithstanding these reasons, the literature also shows that the effectiveness of TQM in HEIs is context-specific and depends on the way the institution goes about TQM to address the issues raised by Zabadi (2013). Thus, there are several accounts in the developing countries, particularly in Middle Eastern and South Pacific jurisdictions, which have successful TQM models in universities (Unal 2001; Arif, Zaidin and Sulong 2007; Becket and Brookes 2008). When TQM is targeted as a strategic management philosophy and practice that embeds service quality and continuous improvement, HEIs are capable of delivering relatively positive results for transformational change.

Developing Higher Education Quality Management Systems

As explained previously, TQM is a long-term management philosophy and practice that focuses on customers' needs, embraces employees in customer service management and supports continuous improvement of products and processes to ensure customer satisfaction. The literature is replete with cases of where organizations have adopted TQM as their management philosophy for transformational change (Dobrzanski and Roszak 2007; Riley et al. 2010; Barry 2012). Dale, van der Wiele and van Iwaarden (2007) show that from the 1940s to now, organizations have been shifting their conceptual approaches and models from inspection to TQM as demonstrated on the next page:

- Inspection
  (measuring products for defaults and appraising services)
- Quality control ↓
  (detecting, measuring and fixing faults of products and services)
- Quality assurance ↓
  (detecting root causes in product and service faults and planning for prevention) ↓
- Total quality management
  (analysing organizational culture, systems and practices for continuous improvement)

Despite the shifts made over the past seventy years, Dale notes that some organizations have not yet moved beyond a quality control or quality assurance approach. This may be a result of their unfamiliarity with quality management approaches or lack of organizational capacity to make the transition.

In the United States, continuous improvement approaches in higher education grew in the late 1980s and early 1990s with the advent of TQM systems such as the ISO 9001 and the Malcolm Baldridge National Quality Award (Dew and Nearing 2004). Focusing on continuous improvement, many public and private universities and colleges started to make the transition from a quality assurance approach to a TQM model. Other models such as Deming's Plan–Do–Check–Act, balanced scorecard methodology and institutional accreditation have adopted TQM principles and approaches. They provide useful tools for organizational transformation. All of these models support Deming's fourteen points for transformation shown in figure 16.1 (www.leadershiptransformationgroup.org). They represent a useful foundation for TQM in all organizations, including HEIs.

Deming's fourteen points encourage the establishment of an organizational leadership and management culture to continuously support both system and cultural change. The focus is on managing transformational change through leading and developing the workforce, managing outputs and outcomes, attaining goals and continuously improving systems. While this chapter explores TQM models with a focus on services, it emphasizes three key issues raised by Deming's transformational change principles. It identifies the important role of leadership for transformation; it discusses how necessary it is for the HEI to undergo continuous process improvement as the basis of its transformational activities and it centres its approach to TQM on workforce education and training.

As noted from Deming's points, organizational learning, by means of double-loop learning approaches, is integral to an effective TQM model. Learning requires commitment from leadership, it must be properly supported and organized and it must be focused on improving the way employees perform to satisfy customers' (including stakeholders') needs. According to Jensen (2010), a HEI is a knowledge organization,

Category	Point
Leadership	1. Institute leadership across the organization
	2. Drive out fear among employees
	3. Remove pride as a barrier to productivity
	4. Break down barriers between staff areas
Management	5. Adopt a new management philosophy
	6. Eliminate micromanagement of employees
Performance	7. Eliminate slogans, exhortations and targets for workforce
	8. Eliminate unnecessary numerical quotes
	9. End the practice of assuming "cheaper is better"
Change and improvement	10. Create constancy of organization's purpose towards service improvement
	11. Continuously improve systems of production and service
	12. Take actions to affect transformation
Education and training	13. Institute vigorous programme of education and training
	14. Institute training as regular part of work life

Figure 16.1. Deming's fourteen points for organizational transformation

Source: Adapted from http://www.leadershiptransformationgroup.com/LTGcampus/resources/understandingtotalqualitymanagement.pdf.

so learning and development is its core business. Faculty who teach and assess learners are critically aware, reflective practitioners and, therefore, they continuously assess the learning environment to bring about improvement in the classroom (Tan 2008). Thus, for knowledge transfer to students and society, the institution, as a whole, must continually assess its performance and encourage employees' learning so that the organization consistently meets its customers' changing needs. This relationship is revealed in figure 16.2, which represents an organizational transformation model that articulates how organizational learning and performance are connected to continuous process improvement and, ultimately, stakeholder impact. This model complements the nine-stage conceptual model for TQM in education by Unal (2001), which includes top management decisions to embrace TQM, collecting customer data to assess organizational performance, educating and training employees, piloting the TQM system, expanding TQM and improving processes to satisfy customers.

Quality Assurance in Caribbean Higher Education

In the Caribbean, HEIs have traditionally developed quality assurance models that focused on academic courses and programmes. Arising out of the 1994 Chancellor's Commission Report of the University of the West Indies (UWI), in 1998, Sir Allister McIntyre, then vice chancellor, was noted to have articulated a quality assurance

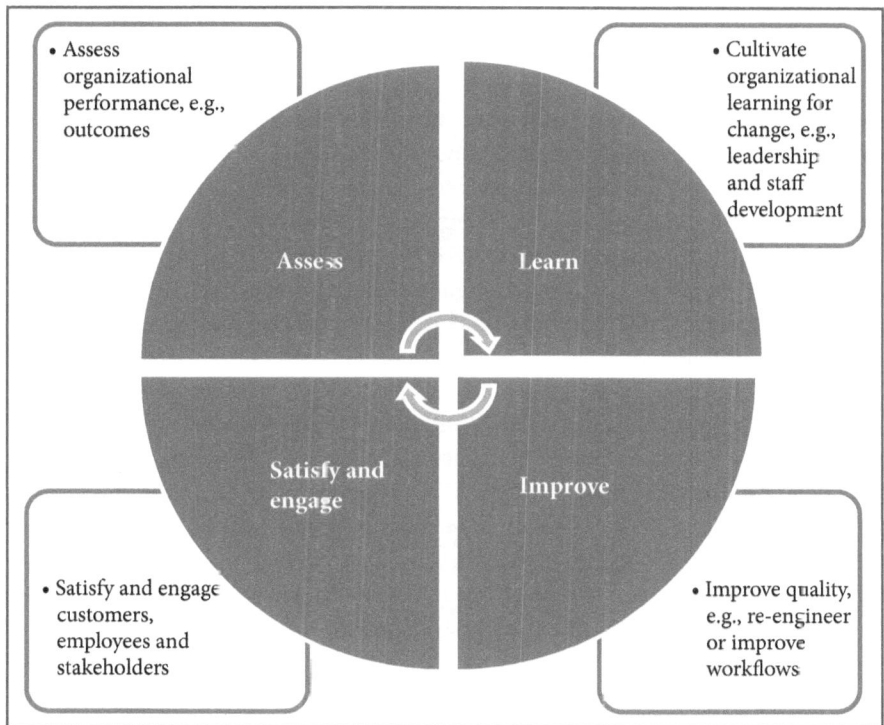

Figure 16.2. Suggested organizational transformation model for higher education
Source: Ali (2012).

model using what was described as the "Hub and Spokes", with UWI as the hub for academic programme quality and tertiary level institutions being the spokes for programme quality delivery. The report recommended the formalization of a quality assurance unit to manage programme quality assurance across the UWI system. The UWI quality assurance model is a framework for systematic academic departmental audits (now called evaluations) and reviews of academic programmes that determine levels of stakeholder satisfaction for programmes and recommend quality improvement measures. This approach to quality assurance, which is a model that has been practised by HEIs in developed countries such as the United Kingdom from as early as the 1980s and 1990s, has been extensively adopted in the Caribbean region till today. HEIs, specialized and professional accreditation bodies and some national agencies such as the University Council of Jamaica have accepted this model for programme quality assurance and this has for many years become the established convention.

Since the mid-2000s, national accreditation agencies in the anglophone Caribbean began to be established. From then, the practices in accreditation have shifted from programme accreditation to institutional registration (a pre-accreditation model) and accreditation. A shift in national accreditation policies began to take HEIs along an alternative pathway, which tended to emphasize the benefits of TQM models for institutional transformation. In fact, both the Malcolm Baldridge performance excellence/quality award and the ISO 9001–2000 (education) models were being adopted as references and standards for developing quality management systems (QMSs) for institutional registration and accreditation. With the advent of compulsory QMSs for higher education in countries such as Trinidad and Tobago, HEIs began to design their internal QMS using the standards espoused by national accreditation agencies. For instance, compulsory registration of HEIs in Trinidad and Tobago requires the adoption of Criterion 3: Quality Management System, which expects that an HEI should have "a well-planned QMS . . . to assure quality of educational outcomes" (http://www.actt.org.tt/index/php/services/registration).

A few Caribbean universities and colleges have since undertaken to develop their QMS. In 2008, adopting the Malcolm Baldridge performance excellence/quality award, the UWI Cave Hill campus developed the first campus QMS (UWI 2008). The system comprised seven organizational processes to include: (1) leadership and change; (2) strategic planning and improvement; (3) customer, stakeholder and market focus; (4) knowledge management, measurement and analysis; (5) faculty and staff management and development; (6) educational and business process management and (7) educational and business results-based management. At about the same time, the University of Trinidad and Tobago developed an organizational intelligence model that uses feedback from unit quality assurance and improvement processes (http://u.tt/index.php?accreditation=1). The organizational intelligence model adopted a knowledge management approach to decision making and process improvement. University of Trinidad and Tobago systematically gathers data, mines the data, assesses performance and evaluates outcomes. Similarly, UWI has embraced balanced scorecard methodology for analysing and reporting its performance in relation its goals in its 2012–17 strategic plan. Continuous improvement is being determined by measuring the attainment of quality outcomes, and the associated root causes and risks to achieving them, with quality tools. Responding to institutional accreditation reviewers' recommendations, the UWI St Augustine campus has developed a system for management of quality that focuses on a continuous improvement strategy with customer service satisfaction approaches. Given the complexity of UWI, this system marries two complementary quality systems, one which is a regional academic quality assurance model and the other a campus-based non-academic QMS that addresses service quality.

Service Quality Models in Higher Education

Baca (2006) highlighted that there is an important relationship between service providers and customers that supports service quality in HEIs. Baca states: "When a service transaction occurs between the service provider and a customer there are dimensions of that transaction that are essential to making the customer feel satisfied with the transaction" (iii).

For a university, service quality is critical to its repeat business success and continuous improvement. Apart from acquiring their knowledge and skills when engaged in course and programme learning, students, in particular, are interested in a range of service transactions offered across several service units, such as admissions services, financial services, information services, product services, student development services, facility services and the list goes on. Jusoh et al. (2004) reported that service quality is a strategic issue that requires continuous measurement of service expectations. They have identified two broad approaches to measuring service quality. SERVQUAL, which was designed by Parasuraman et al. in 1985, is a satisfaction measurement approach that looks at the customer's perceptions against his or her expectations (Naik. Gantasala and Prabhakar 2010). SERVPERF, on the other hand, measures performance of services against a standard. As described by Jusoh et al. (2004), there are essentially eleven service quality dimensions that can be measured in higher education establishments. These dimensions are: (1) reliability; (2) responsiveness; (3) customization; (4) credibility; (5) competence; (6) access; (7) security; (8) communication; (9) tangibles and (10) understanding customers.

While some institutions may implement service quality models in isolation, it is best established within the context of a TQM approach (Becket and Brookes 2008). The TQM approach creates a framework within which service quality measurements can get incorporated by leadership in management decision making and accountability actions. In other words, it provides improvement data and information for planning and management of organization-wide transformational changes. A good example was provided by Canic and McCarthy (2000), who showed that Indiana University Southeast's service quality agenda was linked to planning, leadership and service improvement. They identified a vision for leadership, quality management philosophy, ten-step service quality improvement plan involving staff training and a campus-wide recognition system. The Indiana University Southeast example makes clear that a university service quality system, whether designed within a QMS or stand-alone, should be focused, specific, rational and contextual. This author proposes that an institution may adopt ten steps for devising and implementing a service quality system:

1. develop a QMS that includes services
2. define service quality and develop a philosophy of service quality
3. construct a service quality system
4. develop employee-based measurable standards for service quality
5. train leadership to lead and manage services
6. train and monitor employees' performance against the standards
7. develop and administer service-level agreements for service quality

8. measure customers' expectations of service quality
9. evaluate the service quality system
10. evaluate the QMS within which service quality operates

In an attempt to rationalize service quality in Caribbean HEIs, these steps were considered in the conceptual design of a Caribbean-wide research project focused on improving service quality in Caribbean higher education.

The Caribbean Higher Education Service Quality Research Project

While HEIs have regularly conducted customer service training for line managers and service employees, service quality was first programmed as a staff development agenda at the UWI Cave Hill campus in 2008 by a consultant. The service quality programme was focused on employee productivity. Service standards were established for service employees, leaders were offered a brief orientation to the service quality programme and service employees were trained in mystery shopping to ascertain service areas for improvement. The approach used was based on experiences that were essentially gained from the tourism sector in Barbados, which were not necessarily aligned to the culture and practices of HEIs. Consequently, there were several issues that required critical attention; these can be summed up in a few related questions: How did the employee training model fit within a QMS? How did one determine where to build capacity within a quality system context? Were all units equal or were some of more strategic value than others? Did students, as the main customers, define their vision and perspectives of service quality? Despite the successes gained, these represented some of the challenges as the QMS was being developed at the campus.

In order to interrogate some of these issues relating to service quality in Caribbean higher education, in September 2011, this author under the aegis of the Caribbean Agency for Higher Education Development initiated a three-phased Caribbean Higher Education Service Quality Research Project (Ali 2012). Ali (2010) identified five service categories in a HEI: (1) instructional; (2) learning support; (3) student development; (4) business and (5) research and/or consulting. These service categories vary in function and complexity from one institution to another but represent key transactions for the institution's performance. Nowhere in the Caribbean was there any documentation to ascertain what were the HEI service categories requiring quality improvement. Noting these issues, the Caribbean Higher Education Service Quality project was conceptualized to get a better understanding of service quality practices through the lens of the main customers, the students, and to appreciate what students felt should be the service category priorities and standards to inform service quality measurement models and to enhance service quality in Caribbean higher education.

Three institutions, one a regional public university, another being a regional private university and the other, a regional publicly funded vocational training institution, participated in the Caribbean Higher Education Service Quality study. Forty students from the three HEIs, who represented a random mix of demographics, such as gender, age, religious affiliation, socio-economic, educational, performance and career aspiration levels, were selected by HEI administrators. Students were organized into three focus groups and, with moderated support, were led to: (1) define service quality within their educational setting; (2) analyse the service categories within their institutions and (3) prioritize these service categories. In the end, focus groups presented and discussed their positions. They gave their overall conceptual definition of service quality and categorized service quality areas that were important to their institution. All students agreed to a definition of service quality as: "A system that enables the fulfilment of students' (and clients') needs and expectations during business and educational interactions/transactions and, which ultimately supports the development of the institution's customers (students) as holistic and competent professionals and citizens who are committed to continuous employee development and personal productivity and improvement in the world."

Students were primarily concerned with their professional and personal development and felt that if services were to be measured and improved, it would be best for the HEI to concentrate on providing services that gave students a competitive advantage in terms of their access to and completion of their educational programmes.

When looking at the issue of prioritizing the service categories, with the exception of the regional public university, all students felt that the emphasis of customer services should be directed to their learning support, instruction and student development experience. Students from the regional public university, however, tended to focus on instructional services and research and consulting services. Interestingly, the business transactions such as financial, enrolment, graduation and other administrative support services were not as high a priority. When asked, students explained that this was due to either other options being available for the desired service or because of the minimal and periodic encounters they have had with these services. It could be interpreted that there are possibly two broad service categories: (1) essential services (instructional, learning support and student development) and (2) desirable or convenient services (business, research and consulting). What the study demonstrated is that while service quality is necessary for all services in a HEI, there are two issues to note. A "one-size fits all" service quality model will not work. Service quality is contextual and the model should vary from one HEI to another. Secondly, service quality is most important for services that involve transactions that directly impact students' learning and development experience. This could mean that in order to improve service performance across the institution, strategic focus in the quality system should be placed on essential services that cater to students' academic and holistic development first and then desirable services such as administrative processes after.

The University Service Excellence Approach

The experiences from the Cave Hill campus's service quality programme and Caribbean Higher Education Service Quality research project have helped in developing and establishing a sustainable service excellence strategy that is coherent and articulated to a campus-based QMS at the UWI St Augustine campus. The service excellence strategy has adopted the ten steps proposed by this author and has adapted the eleven service quality dimensions as employee-driven standards for measurement of service excellence. In 2012, the service excellence strategy was developed and launched during the 2012–13 academic cycle as a key component project within an institutional effectiveness programme. As a project, the service excellence strategy initiative placed all service units within essential services (student affairs, library, faculties and student development) and desirable services (business and research/consulting services) categories. Essential services also include the student government organization. While all units are being prepared for their assessment of performance against the service excellence standards, emphasis has been placed on the essential services category. The service excellence strategy also involved the development of an enterprise-wide proprietary web-based service excellence monitoring software system. This software is used for reporting unit service excellence assessment data which can be viewed by decision makers across the campus to ascertain where customer service improvements are needed.

Developing Transformational Leaders and Service Staff

Two critical aspects to organizational transformation in HEIs that should be considered are the development of transformational leaders and facilitation of an organizational learning culture. In examining the influence of transformational leadership on TQM within Libyan higher education, Argia and Ismail (2013) argue that transformational leadership can only transform employees within a university if they are effectively led to learn new tools and methods, and improve how they perform. In reflecting upon transformational leadership within UWI, Howe and Newton (2000) posited that transformational leadership is frustrated by the bureaucratic mode of operation, particularly to support effectiveness in service delivery, teaching and learning. This issue would require not only new structures and systems for change but also leaders and employees who are committed to and embrace transformational change through their own work-based learning. It was for this reason that the initial focus of the institutional effectiveness programme was on developing the transformational leadership capacity of teams of service leaders for managing and enhancing customer services excellence in each unit. The project was initiated by the campus principal with the support of the campus's leadership teams. The campus senior management team endorsed the policy framework, the extended management teams supported it and the campus's academic governing organ approved it. A very important aspect of the initiative is leadership development training. Leadership teams are trained to

- develop new markets and products for building new customers and customer loyalty;
- access, analyse and use institutional research data for customer services improvement;
- use customer profiles to create customer services strategies;
- develop customer complaints policies; and
- manage service amenities.

The ultimate goal of the sessions was to enable leaders to make empirically sound decisions, plans and policies, improve key value-added processes (workflows) and determine outcomes for organizational units to measure customer service excellence. Three cohorts of unit service leaders from twenty-seven non-academic units were trained in February 2013 to manage service excellence with more cohorts to be trained in the current academic cycle. Overall, 97 per cent of the leaders trained were satisfied that the training impacted their knowledge of service excellence and provided the methods and digital tools for effectively measuring it. Of the twenty-seven non-academic and faculty units trained, six months later, fourteen of these units have produced or are developing documentation of their service excellence practices.

While leadership capacity for service excellence is critical, service employees have to interface with a new quality system. Apart from developing the skills needed for engaging customers in a dynamic service quality system, administrative, technical and service support employees have to acquire the attitudes and behaviours necessary to ensure that the standards are being delivered. It is these employees who constantly and directly interact with customers – their needs, transactions, complaints – and other employees, who as part of a team, deliver the goods and services customers require. Customer service excellence training was also packaged as a module within existing customer service courses offered to employees within UWI. This provides for sustainability in employee development. The service excellence strategy initiative is presently entering the second year of its work programme. Apart from its transformational change capacity at the St Augustine campus, it has already attracted attention across other campuses and been noted as a major initiative in the institutional effectiveness programme for the UWI system.

Conclusion

TQM, as an organizational transformation model for HEIs, provides the management philosophy and practices for continuous improvement. In the Caribbean, the TQM experience has emerged from HEIs' self-reflection on their effectiveness models, international approaches to higher education management and institutional registration and accreditation policies and practices. Higher education leadership and organizational learning are two critical aspects to transformational changes within the TQM framework. If a contextualized and suitable service quality system is developed and managed within a TQM model and, if this system focuses on service quality leadership and employee team building, the results can be truly transformational. In spite of cultural change issues, the UWI service excellence experience bears testimony to this point. Leaders felt that they experienced transformational change through service quality leadership development and are presently putting measures in place to develop service excellence standards across the campus. Service employees are routinely being exposed to new and emerging workplace practices for modifying their competencies in a customer-focused organization. The ultimate goal is having highly satisfied employees who continuously delight their customers and, in turn, bring greater accountability and satisfaction to shareholders/stakeholders.

References

Agasisti, T., and F. Bonomi. 2013. "Benchmarking Universities' Efficiency Indicators in the Presence of Internal Heterogeneity". *Studies in Higher Education* 1–19. doi:10.1080/0307 5079.2013.801423.

Ali, E. 2010. *Position Paper on Governance and Management of Service Quality*. Cave Hill, Barbados: Quality Assurance Office, UWI.

———. 2012. "The Caribbean Higher Education Service Quality Project: An Empirical Study of Three Caribbean Higher Education Institutions". Paper presented at the tenth annual conference of the Caribbean Area Network for Quality Assurance in Tertiary Education, Montego Bay, Jamaica.

Al Tasseh, G.H. 2013. "Obstacles to the Application of Total Quality Management (TQM) in Higher Education Institutions in the State of Kuwait". *European Scientific Journal* 9 (4): 209–20.

Argia, H.A.A., and A. Ismail. 2013. "The Influence of Transformational Leadership on the Level of TQM Implementation in the Higher Education Sector". *Higher Education Studies* 3 (1): 136–46.

Arif, M.S.M., N. Zaidin and N. Sulong. 2007. "Total Quality Management Implementation in Higher Education: Concerns and Challenges Faced by the Faculty". *Best Practices in Education and Public Administration* 7 (3): 1-23.

Baca, D.R. 2006. "Dimensions of Service Quality of the University of Arizona Sponsored Projects Offices Services Internal Customers". PhD diss., Texas A&M University, Texas.

Barry, W.J. 2012. "Challenging the Status Quo Meaning of Educational Quality: Introducing Transformational Quality (TQ)". *Educational Journal of Living Theories* 5 (1): 1-26.

Becket, N., and M. Brookes. 2008. "Quality Management Practice in Higher Education: What Quality Are We Actually Enhancing?" *Journal of Hospitality, Leisure, Sport and Tourism Education* 7 (1): 40-54.

Canic, M.J., and P.M. McCarthy. 2000. "Service Quality and Higher Education Do Mix: A Case Study Exploring the Service Environment at Indiana University Southeast". *Quality Progress* 33 (9): 41-46.

Dale, Barrie G., Ton van der Wiele and Jos van Iwaarden. 2007. *Managing Quality*. 5th ed. Malden, MA: Wiley-Blackwell.

Dew, J.R., and M.M. Nearing. 2004. "Embarking on Continuous Improvements in the Academic Community". In *Continuous Quality Improvement in Higher Education*, edited by J.R. Dew and M.M. Nearing, 15-32. Westport, CT: American Council of Education/Praeger.

Dobrzanski, L.A., and M.T. Roszak. 2007. "Quality Management in University Education". *Journal of Achievements in Materials and Manufacturing Engineering* 24 (2): 223-26.

Dounos, P.K., and G.A. Bohoris. 2007. "Exploring the Interconnection of Known TQM Process Improvement Initiatives in Higher Education with Key CMMI Concepts". Paper presented to the tenth QMOD conference. Lund University, Sweden.

Hogg, R.V., and M.C. Hogg. 1995. "Continuous Quality Improvement in Higher Education". *International Statistical Review* 63 (1): 35-48.

Howe, G.D., and E. Newton. 2000. "A Vision for Transformational Leadership for the University of the West Indies". In *Higher Education in the Caribbean Past, Present and Future Directions*, edited by G.D. Howe. Kingston: University of the West Indies Press.

Jensen, H.S. 2010. "Working Paper 14: The Organization of the University". In *Working Papers on University Reform*. Aarhus, Denmark: Danish School of Education, University of Aarhus.

Jusoh, A., S.Z. Omain, N.A. Majid, H.M. Som and A.S. Shamsuddin. 2004. *Service Quality in Higher Education: Management Students' Perspective*. Malaysia: Research Management Centre, University of Technology.

O'Donoghue, T., and S. Clarke. 2010. "Organisational Learning and the Intelligent School". In *Leading Learning: Processes, Themes and Issues in International Contexts*, edited by T. O'Donoghue and S. Clarke, 110-14, 162-82. London: Routledge.

Mainardes, E.W., H. Alves and M. Raposo. 2011. "The Process of Change in University Management: From the 'Ivory Tower' to 'Entrepreneurialism' ". *Transylvanian Review of Administrative Sciences* 32 (E): 124-49.

Morgan, G. 2006. "Learning and Self-Organization: Organizations as Brains". In *Images of Organization*, edited by G. Morgan, 71-114. Thousand Oaks, CA: Sage.

Naik, C.N.K., S.B. Gantasala and G.V. Prabhakar. 2010. "Service Quality and Its Effect on Customer Satisfaction in Retailing". *European Journal of Social Sciences* 16: 239-51.

Neefe, D.O. 2001. "Comparing Levels of Organizational Learning Maturity of Colleges and Universities Participating in Traditional and Non-Traditional (Academic Quality Improvement Project) Accreditation Processes". MSc research paper, University of Wisconsin-Stout.

Pedder, D., and J. MacBeath. 2008. "Organisational Learning Approaches to School Leadership and Management: Teachers' Values and Perceptions of Practice". *School for Effectiveness and School Improvement: An International Journal of Research, Policy and Practice* 19 (2): 207–24. doi:10.1080/09243450802047899.

Pineda, A.P.M. 2013. "Total Quality Management in Educational Institutions: Influences on Customer Satisfaction". *Asian Journal of Management Sciences and Education* 2 (3): 31–46.

Riley, W.J., H.M. Parsons, G.L. Duffy, J.W. Moran and B. Henry. 2010. "Realizing Transformational Change through Quality Improvement in Public Health". *Journal of Public Health Management and Practice* 16 (1): 72–78.

Tan, C. 2008. "Improving Schools through Reflection of Teachers: Lessons from Singapore". *School for Effectiveness and School Improvement: An International Journal of Research, Policy and Practice* 19 (2): 225–38. doi:10.1080/09243450802047931.

Tossey, P. 2003. "The Learning Organisation". In *The Theory and Practice of Learning*, edited by P. Jarvis, J. Holford and C. Griffin, 144–56, 171–84, 192–93. London: Kogan.

Unal, O.F. 2001. "Application of Total Quality Management in Higher Education Institutions". *Journal of Qafzal University* 7 (2): 1–18.

UWI (University of the West Indies). 2008. "Cave Hill Campus Quality Management System". Academic Board, Quality Assurance Office, Cave Hill, Barbados.

Zabadi, A.M.A. 2013. "Implementing Total Quality Management (TQM) on Higher Education Institutions – A Conceptual Model". *Journal of Economics and Finance* 1 (1): 42–60.

17 | Future Directions for Quality Assurance in Higher Education

ANNA KASAFI PERKINS

"There are new challenges that come up all the time."
– Stamenka Uvalic-Trumbic, Consultant,
CHEA International Quality Group (Fischer 2012)

Higher education is a diverse and dynamic sector, so ensuring and enhancing quality will itself be a diverse and dynamic process. This will require quality assurance practitioners keeping abreast of institutional, national, regional and international trends – what former pro vice chancellor and chair for the Board for Undergraduate Studies at the University of the West Indies (UWI) Alvin Wint often referred to as "environmental scanning and mapping". It is important to take these trends seriously as the Caribbean is impacted by developments such as the debate about global quality standards (Fischer 2014) and entrenching enhancement in the quality process as is happening in Scotland (QAA Scotland 2012); at the same time, Caribbean quality assurance practitioners themselves contribute to shaping the regional and global trends by their own activities; indeed, institutions with fairly mature and robust internal quality assurance systems like the University of the West Indies and the University of Trinidad and Tobago are on a trajectory to move beyond assurance towards enhancement. Such enhancement is evident in the formalized linkage between quality assurance and strategic planning, as detailed in the Scottish higher education experience: "Quality processes are becoming less retrospective and more evaluative, more focused on forward planning and more effectively linked to strategic planning processes. The evidence suggests they are also becoming more student-centred, both in terms of student involvement in decision-making processes, and in terms of a holistic strategic focus on the continuous improvement of the student learning experience" (Barron 2007).

Both universities clearly tie in strategic planning and quality assurance with evidence-based decision making (see, for example, http://u.tt/index.php?accreditation=1); deepened student engagement in quality assurance processes such as surveys and governance is also a part of improving student learning. The challenge will be for the external quality assurance agencies (EQAAs) to ensure that their processes take account of enhancement thinking and practices.

Exciting Developments

In the few months between penning the introduction to this volume and writing this final chapter, all kinds of exciting developments have been evident in the region: the University College Cayman Islands (UCCI) signed a memorandum of understanding with the Chartered Institute of Legal Executives (CILEx); an excellent volume entitled *Education in the Commonwealth Caribbean and the Netherlands Antilles* (Thomas 2014) was published; the University of Trinidad and Tobago has partnered with online course provider Coursera to establish a "network" of online courses that will be accessible throughout the country, the government of Jamaica has started to train aspiring principals for the classroom in recognition that leadership is the key factor in quality education, the Second Caribbean Conference on Higher Education in the Caribbean hosted by UNESCO–International Institute for Higher Education in Latin America and the Caribbean and UWI was held (best practices in higher education was the focus), and Caribbean Area Network for Quality Assurance in Tertiary Education has published its scoping study of quality assurance provisions across the Caribbean, including Suriname. These local and regional developments all point to changes in thinking about and practice of monitoring, assuring and enhancing provisions in higher education.

Global Accreditation Standards

Increased international and cross-border collaborations such as the UCCI–CILEx joint degree lead to concerns about creative ways of accrediting such joint degrees and the meaning of accreditation by professional bodies such as CILEx. The UCCI–CILEx memorandum of understanding is the result of a unique collaboration, which allows CILEx, a UK-based professional organization, to award credentials jointly with UCCI. The collaboration also enables UCCI graduates to further their professional development by transferring credits to CILEx's professional programmes (UCCI website). This collaboration is evidence of the value of institutional prestige and branding as part of the process of attesting to quality. With increasing numbers of students participating in cross-border education, there are growing concerns with quality and comparability. The response has been a burgeoning conversation on global standards for cross-border education. Some argue that the absence of such global accreditation standards as well as the opaqueness of accreditation processes has led to international "rankings becoming a de facto quality assurance system" (Fischer 2014). Of course, creating a global accreditation standard will not be easily done given the diversity of institutions and the dynamism inherent in the higher education sector – one size would not and should not fit all. Students would certainly benefit by being able to compare quality of education across borders. Institutions would also benefit in terms of benchmarking and the advantages that could accrue from cross-border collaborations and partnerships.

Massive Open Online Courses and Cross-Border Offerings

One of the first responses to question of global accreditation standards has been the formation of the Council of Higher Education (CHEA) International Quality Group. This membership group was formed in 2012 to "serve as a forum for issues of international accreditation and quality assurance for the regulation of overseas branch campuses and the oversight of online courses" (Fischer 2012). Council of Higher Education is an American group of colleges and accrediting bodies and is, perhaps understandably, in the forefront of responses to the challenges of both ethics and quality that have surfaced as American colleges have established overseas campuses; the experience of providing cross-border education has led some colleges to question the quality of local accreditation bodies or to be unclear about the standards by which their own accrediting bodies would be assessing their overseas programmes (Fischer 2012).

Similarly, the massive open online courses phenomenon, which had its genesis in American higher education institutions, continues to pose questions of quality and validity that can no longer be ignored, as is evident in the UWI's move to draft a policy concerning massive open online courses and University of Trinidad and Tobago's collaboration with Coursera (Mead 2014). The latter partnership is Coursera's first national network, and will provide online courses, learning materials and television programmes. There will be learning centres on the University of Trinidad and Tobago campuses, where students will be able to get help from mentors. The online course will also be accredited and students can earn college credits. By whom and how is as yet unclear.

Tackling Grade Inflation

The Thomas collection (*Education in the Commonwealth Caribbean and the Netherlands Antilles*) is interesting for what it does not do – give particular attention to matters of quality assurance in higher education. Most of the chapters continue the focus on primary and secondary education across the region, while paying little attention to higher education. Nonetheless, some issues of quality assurance in higher education were evident, for example, concerns with grade inflation in courses and examinations at the University of Belize, which are addressed through accreditation and partnership targets. Internal and external quality assurance processes are also spoken of, but with little detail (Thomas 2014). Nonetheless, the question of grade inflation is a live and ongoing one.

In the United Kingdom, a unique experiment is being run to tackle exactly the question of grade inflation. This grade inflation is said to be evident in the number of graduates being awarded first-class and upper-second degrees (2.1). Twenty-one universities have been experimenting with American-style grade point average (GPA) in the hope that it will be a better evaluation of student performance and help solve the problem of grade inflation, among others (Hodges 2014). Instituting such a system is expected to allow universities to keep better tabs on rising grades over time. In fall 2014, the experimenting universities will provide students with both the traditional degree class and GPA as well as publish a report on student performance. Critics of the experiment point to several problems with the experiment, however. Confusion of terminology is one potential problem as in the British system, for example, 2.1 already has a particular meaning. Some do not believe that it will address the problem of degree inflation (currently two-thirds of British students obtain a first class honours degree). Others argue that the problem of grade inflation has yet to be established and perhaps the bigger conversation should be on marking practices.

UWI has pioneered for many years with a hybrid system of honours degrees married to a modified GPA system, a study of which may prove useful as the United Kingdom wrestles with its system. The UWI's adoption of GPA is intended to allow graduates to link more easily with programmes at mainly North American universities, where the GPA system is entrenched. Performance of UWI students can be easily understood by institutions that rely on this method. At the same time, using GPA as a mechanism for calculating honours ensures that student performance is transparent to Commonwealth and other universities that maintain honours system. The UWI GPA system was modified as recently as 2014.

Accountability for EQAAs

The Caribbean Area Network for Quality Assurance in Tertiary Education Scoping Study was conducted against the background of the 2002 decision made by Caribbean Community governments to set up autonomous national accreditation bodies as part of a network of External Quality Assurance Agencies "expected to develop common standards and measures of quality assurance and accreditation" (CANQATE 2014). Among the findings of the study is that the Caribbean

Community Model for the EQAAs may have gone too far in proposing a range of functions for national EQAAs that was not feasible for all. A further area of concern is the accountability mechanisms for the Caribbean EQAAs. The study establishes that none of the EQAAs surveyed has in place a functioning accountability process, even the oldest, the University Council of Jamaica, which predates the Caribbean Community Model. Indeed, Perkins and Brown (2014) assess the University Council of Jamaica in light of the standards of good practice established by International Network of Quality Assurance Agencies in Higher Education and find that, since its formation in 1987, the council has not itself undertaken an institutional review of its processes and performance (interestingly, the survey describes the University Council of Jamaica process as "developed and partially implemented" [7]). Similarly, the Accreditation Council of Trinidad and Tobago mechanisms are described as "developed and partially implemented" (9). No indication is given as to what these mechanisms are, however.

Clearly, there is a need for the process of "accrediting the accreditors" to be given priority status as the matter of quality assurance in higher education in the region is tackled. The long-stalled Caribbean Community Accreditation Agency for Education and Training is the regional mechanism, which, once fully functioning, would assist the EQAAs in carrying out their functions through developing core criteria, standards of good practice and procedures. There is perhaps an urgency here for this mechanism to be established to serve as a means of accountability for EQAAs, especially through providing a regional system for continuous quality improvement of their performance. Clearly, a set of guidelines would need to be developed for the accountability procedures of the EQAAs, including a published policy for assuring the quality of the work of the agency, mandatory cyclical external review of the agencies' activities at least once every five to seven years, and an internal process of quality assurance with feedback mechanisms.

In conclusion, Caribbean EQAAs, internal quality assurance offices and higher educational institutions should consider ways in which they can respond to and benefit from the varying developments within the sector. At the same time, they should undertake institutional research into their own practice and contributions to the global trends, especially with an eye to new models of quality assurance that will tackle issues such as grade inflation and accreditation. It is clear that the quality assurance issues that are tackled are by no means unique to the region, so global solutions must be foregrounded.

References

Barron, Thelma. 2007. "Managing Assurance and Enhancement in the Scottish HE Sector: Evolution and Convergence". *Higher Quality (Bulletin of the Quality Assurance Agency for Higher Education)*, no. 23 (March).

CANQATE (Caribbean Area Network for Quality Assurance in Tertiary Education). 2014. "Scoping Study of Quality Assurance in Tertiary Education in the Caribbean: Executive Study".

Fischer, Karin. 2012. "New Group Serves as Forum for Global Academic Quality Issues". *Chronicle of Higher Education* (online), September 13.

———. 2014. "Accreditors Debate the Merits – and Feasibility – of Global Standards". *Chronicle of Higher Education* (online), January 31.

Hodges, Lucy. 2014. "British Experiments with GPA, but Critics Say It Won't Be the Cure All". *Chronicle of Higher Education* (online), January 27.

Mead, Russell, and staff. 2014. "Education Innovation National MOOCs in the Caribbean". *American Interest*, May 31. http://www.the-american-interest.com/blog/2014/05/31/national-moocs-in-the-caribbean/.

Perkins, Anna Kasafi, and Denise Brown. Forthcoming. "Quality in Higher Education a Decade after the Education Taskforce Report: Evaluating the Performance of the UCJ". Submitted for publication.

QAA Scotland (Quality Assurance Agency). 2012. *Handbook for Enhancement-Led Institutional Review: Scotland*, 3rd ed. QAA 033 0412003. http://www.comms.qaaac.uk/reviews/ELIR/handbook/scottish_ hbook _prefuce.asp.

Thomas, Emel. 2014. "Belize: Seeking Quality Education for National Development". In *Education in the Commonwealth Caribbean and the Netherlands Antilles*, edited by Emel Thomas, 106–19. London: Bloomsbury.

Contributors

Anna Kasafi Perkins is Senior Programme Officer, Quality Assurance Unit, Office of the Board for Undergraduate Studies, University of the West Indies, Vice Chancellery, serving the Mona campus, Jamaica.

Eduardo R. Ali is Chairman of the Caribbean Agency for Higher Education Development, Programme Manager-Institutional Effectiveness, University of West Indies, St Augustine, Trinidad and Tobago, and Vice-President of the Caribbean Area Network for Quality Assurance in Tertiary Education.

Ruby S. Alleyne is Vice-President, Quality Assurance and Institutional Effectiveness, University of Trinidad and Tobago.

Patrick Anglin is Senior Information Technology Officer, Policy and Infrastructure, Office of the University Chief Information Officer, University of the West Indies, Regional Headquarters, Mona, Jamaica.

Dameon A. Black is Deputy President, University College of the Caribbean, Kingston, Jamaica.

Alan Cobley is Professor of History, Pro Vice Chancellor and Chair for the Board for Undergraduate Studies, Vice Chancellery, University of the West Indies.

Patrick S. Dallas is Chairman and Chief Executive Officer, Convergent Technologies Limited, and Chair of the Advisory Committee in Chemical Engineering, University of Technology, Kingston, Jamaica.

Pamela Dottin is Programme Officer, Quality Assurance Unit, University of the West Indies, with responsibility for the Open Campus.

Paulette J. Dunn-Pierre is Chief Executive Officer, Dunn, Pierre, Barnett & Company Limited, a human resource development and consulting firm with offices in Jamaica.

Anna-May Edwards-Henry is Director, Centre for Excellence in Teaching and Learning, University of the West Indies, St Augustine, Trinidad and Tobago.

John Gedeon is Planning Officer, University Office of Planning and Development, University of the West Indies, St Augustine, Trinidad and Tobago.

Sandra Gift is Senior Programme Officer, and Head, Quality Assurance Unit, Vice Chancellery, University of the West Indies, St Augustine, Trinidad and Tobago.

E. Nigel Harris is Vice Chancellor, University of the West Indies, Chairman, Caribbean Examinations Council, and President, Association of Caribbean Universities and Research Institutes.

Halima-Sa'adia Kassim is Senior Planning Officer, University Office of Planning and Development, Vice Chancellery, University of the West Indies, St Augustine, Trinidad and Tobago.

Halden A. Morris is Senior Lecturer, School of Education, University of the West Indies, Mona, Jamaica.

Kofi K. Nkrumah-Young is Vice-President, Planning and Operations, University of Technology, Kingston, Jamaica, and President of the Association of Caribbean Higher Education Administrators.

G. Junior Virgo is Programme Chair (Chemical Engineering), University of Technology, Kingston, Jamaica.

S. Joel Warrican is Director of the Division of Academic Programming and Delivery, University of the West Indies, Open Campus.

June Wheatley is Senior Administrative Assistant, Quality Assurance Unit, Office of the Board for Undergraduate Studies, University of the West Indies, Vice Chancellery, serving the Mona campus, Jamaica.

www.ingramcontent.com/pod-product-compliance
Lightning Source LLC
Chambersburg PA
CBHW031708230426
43668CB00006B/147